Springer Geophysics

The Springer Geophysics series seeks to publish a broad portfolio of scientific books, aiming at researchers, students, and everyone interested in geophysics. The series includes peer-reviewed monographs, edited volumes, textbooks, and conference proceedings. It covers the entire research area including, but not limited to, applied geophysics, computational geophysics, electrical and electromagnetic geophysics, geodesy, geodynamics, geomagnetism, gravity, lithosphere research, paleomagnetism, planetology, tectonophysics, thermal geophysics, and seismology.

More information about this series at http://www.springer.com/series/10173

Fernando Sansò · Mirko Reguzzoni ·
Riccardo Barzaghi

Geodetic Heights

Springer

Fernando Sansò
Department of Civil and Environmental Engineering (DICA)
Politecnico di Milano
Milan, Italy

Mirko Reguzzoni
Department of Civil and Environmental Engineering (DICA)
Politecnico di Milano
Milan, Italy

Riccardo Barzaghi
Department of Civil and Environmental Engineering (DICA)
Politecnico di Milano
Milan, Italy

ISSN 2364-9119 ISSN 2364-9127 (electronic)
Springer Geophysics
ISBN 978-3-030-10453-5 ISBN 978-3-030-10454-2 (eBook)
https://doi.org/10.1007/978-3-030-10454-2

Library of Congress Control Number: 2018967739

This Springer imprint is published by the registered company Springer Nature Switzerland AG
The registered company address is: Gewerbestrasse 11, 6330 Cham, Switzerland

Foreword

Gravity provides the natural orientation in our living environment. The walls of houses are vertical; floors are horizontal disregarding the few avant-garde counterexamples of modern architecture. The two elementary tools are thereby plumb and level. Heights are the central measure in this context. They tell us in which direction water would move, from A to B or from B to A. Points located on one and the same level surface should ideally possess the same value of height. All of this is elementary and rather familiar.

These concepts gain high complexity, however, if they need to be applied on scales beyond our immediate surroundings, when their application refers to a province or even a country. As an example, let us take the control of the water system in a low-land country such as the Netherlands. Large parts of the Netherlands are below sea level. How to ensure that water is managed in a steady, precise and most of all, safe manner using a system of thousands of canals? This reminds us of the pivotal function which all these scenic windmills once had.

Heights and the question of higher and lower gain a scientific dimension when dealing with global height systems and sea level. How to compare heights in Europe with those in America or Australia? Before the background of climate change, it is a great challenge to identify tiny signals of sea-level rise or fall in a dynamic system with winds, surges, ocean tides, plate tectonics and land uplift. Already, more than 120 years ago, some leading European geodesists tried to compare mean sea level at tide gauges of all the seas adjacent to Europe. The comparison of the heights of tide gauges thousands of kilometres apart by means of precision levelling used to be and still is a great challenge. Various disturbances tend to blur the results. In the seventies of the last century, oceanographers and geodesists were arguing about the slope of mean sea level along the east coast of North America. Geodetic levelling showed sea level to slope downwards towards the equator, while oceanographers argued for an upward slope. Only recently with the help of global satellite positioning and modern satellite gravimetry, the controversy could be resolved. Regrettably for geodesists, oceanographers were right. Global mean sea surface, i.e. the ocean surface without exterior influences such as tides, winds and atmospheric pressure, almost coincides with a level surface, with the geoid. The deviations, referred to as sea surface topography, are typically less than one metre with maximum values in the centres of the major ocean current systems. Now, modern geodetic space techniques start to open the door for a worldwide comparison and monitoring of sea level relative to the geoid.

All these rather worldly considerations are concerned with the role of heights in practice and research. But what are the theoretical foundations of heights and height systems? Surprisingly, no book exists so far, covering all relevant elements together and in a concise form. Now, Fernando Sansó, Mirko Reguzzoni and Riccardo Barzaghi, all from the Politecnico di Milano, fill this gap with their book "Geodetic Heights". Chapters 1–5 cover the theoretical foundations: coordinate systems in three dimensions, the theory of the gravity field, gravity quantities relevant for height computation together with the relevant functional models and geodetic coordinates, i.e. coordinate systems related to the Earth's gravity field. Chapters 6 and 7 cover the actual observation models as well as the definition of local and global height systems. Already in the past, the geodesy group of the Politecnico greatly influenced the debate on the theoretical principles of geodetic heights. With the present book, the authors give the reader a comprehensive introduction into the essentials of modern height systems.

Munich, Germany
September 2018

Reiner Rummel
Professor Emeritus
Astronomical and Physical Geodesy
Technical University Munich

Contents

Chapter 1
Introduction

A height is a coordinate in $\mathcal{R}^3$, used in a certain subset of our space, particularly in the framework of physical sciences of the Earth, to discriminate higher from lower points, in some sense to be specifically stated by the type of the height chosen.

So first of all a height, as a coordinate for a subset of points in $\mathcal{R}^3$, is one coordinate in a triple and it makes little sense to define it without specifying the other two. Second, we need to restrict the physical purpose for which a height will be considered in this monograph.

We shall do that by a counterexample and then by stipulating a criterion to identify what kind of "heights" we are interested in, namely geodetic heights. For instance the height of a point on or above the Earth surface could be defined as the air pressure at that point, in a triple completed by two cartographic coordinates, in a specific area of the Earth. Such a concept of height is in fact used in atmospheric sciences to simplify the equations of the dynamics of the atmosphere, and even in common life in mountain excursions. However we rule out this concept, because we know that such a coordinate can significantly change from hour to hour at the same point, fixed with respect to the solid Earth. So we shall agree that we want to study heights that do not change in time, at least they do not change significantly over a time span in which the Earth can be considered as a stationary body.

One could object that the Earth undergoes not only slow geological movements, but also periodical deformations, for instance the body tides due to the attraction of the Sun and the Moon, that are in the range of 1 m and have a main semidiurnal period. Such effects will be considered as perturbations, globally known and subtracted to all the physical quantities considered in this work, so that we shall refer to an idealized static image of the Earth.

After these preliminary remarks, it is time to go the heart of the question, namely in which sense we intend to discriminate higher from lower points. This is primarily related to the gravity field and its direction. Locally, this is first of all related to our physiological sensations. A man standing on the ground defines the direction of the vertical and subsequently a small area under his feet, when they are kept orthogonal

F. Sansò et al., *Geodetic Heights*, Springer Geophysics,
https://doi.org/10.1007/978-3-030-10454-2_1

to the body axis, is horizontal. So a human body is itself a local gravity sensor; certainly not the only one! A short pendulum can be used to define the direction of the vertical at a point in space, also a small quantity of water, still in a bucket will define a horizontal plane at its centre and so forth. The vertical direction, that we shall identify by a unit vector $\boldsymbol{n}(\mathrm{P})$, is in the opposite direction of that of the gravity attraction, and once we know how to materialize it we can connect points that are close one to the other in the vertical direction. In this way starting from point, e.g. on the Earth surface, we can generate a line upward and in fact, assuming we are able to enter into the body of the Earth, also downward. This is a line of the vertical or plumb line; if we do the same at all points in the region of interest, we generate a family of lines, also referred to as the congruence of vertical lines. They have the property that at every point they are tangent to the direction of the vertical. It is a fundamental theorem, consequence of the famous theorem of existence and uniqueness of solutions of ordinary differential systems, that in the region of our interest, where the gravity attraction field never goes to zero and it is at least Hölder continuous, these lines can never intersect, nor even be tangent to one another at any point. In other words, in our region, through any point P there passes one and only one line of the vertical.

Once the congruence of vertical lines is established, one can also consider the family of surfaces that admit plumb lines as orthogonal trajectories. It turns out that these are equipotential surfaces of the gravity field, as we shall see later on, and as such they cannot intersect too. Moreover, always in the range of some kilometers up and down, it happens that the equipotentials are closed surfaces and therefore they are contained one into the other. Plumb lines and equipotentials are the two main ingredients of the geometry of the gravity field, which has been investigated in depth in the 50ies, 60ies and 70ies of the 20th century (Bomford 1952, Marussi 1985, Hotine 1969, Krarup 2006, Heiskanen and Moritz 1967, Grafarend 1975, just to mention a few). We shall use only some of these results, to be presented later on in the text.

We need now to better specify what is the region of our interest, where we want to establish and use geodetic height coordinates. Indeed this region has to include the surface of the Earth S; in particular we want coordinates which are good for all of S, a layer of points above and a layer below it. The reason why we want to cover the whole of S is because in the era of Global Navigation Satellite System (GNSS) measurements we are able to connect any point on the Earth surface by observations that need to be modelled by a unique coordinate system. The reason we limit the region by layers, say of $\pm 30\,\mathrm{km}$ width, is twofold. When we go far away from S in the upper direction, there is a quite substantial change of the geometry of the gravity field. For instance equipotentials at a distance of about 42,000 km from the barycenter become open surfaces (see Sansò and Sideris 2013, Sect. 1.9). This is basically the reason why the potential cannot be used as a height coordinate at least throughout the whole space. So 30 km above S is a good layer for both Geodesy

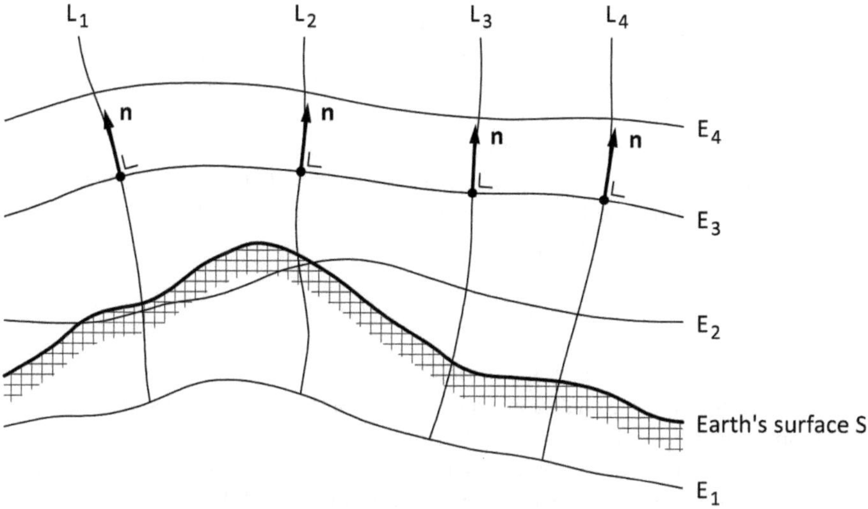

Fig. 1.1 A portion of the Earth surface S with part of the region of interest close to it; equipotential surfaces (E_1 E_2 E_3 E_4) and lines of the vertical (L_1 L_2 L_3 L_4)

and sciences of the lower atmosphere.[1] As for the layer below S one can observe that this depth is sufficient to include important isostatic compensation surfaces, like the Moho, which are traditionally discussed in the framework of Geodesy, as well as the region where most of crustal geophysical phenomena have a seat. When we go deeper we might incur into a more irregular behaviour of the gravity field and at the same time we have an increasingly poor knowledge of the distribution of the masses, which is the origin of the gravity field (Lambeck 1988; Anderson 2007). In the figure below we summarize in a pictorial form what we said above for a portion of our region of interest (Fig. 1.1).

It is worth underlining that given the properties of equipotential, i.e. horizontal, surfaces, they are naturally ordered from below to above, considering that also the plumb lines that they cross orthogonally have a natural positive verse inherited from the vertical unit vector $\boldsymbol{n}(\mathrm{P})$.

We are ready now to give a first definition of what we can consider a geodetic height, that we call here generally as q_3, the third coordinate in a triple (q_1, q_2, q_3). As all regular coordinates, q_3 will have a coordinate line ℓ_3 with a tangent unit vector $\boldsymbol{e}_3(\mathrm{P})$ attached at any point P on it; we say that q_3 is a geodetic height if the relation

$$\boldsymbol{n}(\mathrm{P}) \cdot \boldsymbol{e}_3(\mathrm{P}) \geq k > 0 \tag{1.1}$$

[1] *Note*: indeed Geodesy is also interested in satellite dynamics even for very high satellites, yet at that altitude not all the coordinates discussed in this book are of particular significance: whence the reason to limit our discussion to a bounded region.

is satisfied for all points in the region of interest and for some fixed, positive k. The meaning of (1.1) is that when we move along ℓ_3 in the positive sense, that we call upward, we expect to cross horizontal surfaces which are monotonously above one to the other.

To be more specific, since we do not like to use a coordinate whose lines cross the horizontal surface at a small angle, we could say that we expect that ℓ_3, i.e. $\boldsymbol{e}_3(\mathrm{P})$, is almost orthogonal to the equipotential passing by P. In practice we shall treat situations in which the constant k of (1.1) is

$$k = 1 - \varepsilon \qquad \varepsilon = \mathcal{O}(10^{-2}) \ ,$$

i.e. $\boldsymbol{n}$ and $\boldsymbol{e}_3$ form at most an angle of one or few degrees.

In this monograph we shall consider mainly four types of heights, that we name as orthometric, ellipsoidal, normal and orthonormal (or normal orthometric as they are called in literature), plus some variants. The focus of the book is on two issues: to find the relations between one system of coordinates and the other, which, as we shall see, implies a fine knowledge of the actual gravity field of the Earth; to find the relation between various heights and the quantities that are observable by geodetic techniques.

In this respect a last remark is in order; we often speak of a height system and by that we mean that not only we have a mathematical definition of the coordinate but we have also defined a reference system for it, which is essential to find the connection between this height and the observable quantities. This completely parallels what happens with all types of coordinates in Geodesy. In particular we fix a height system when, given a certain geometry of the coordinate lines and of the corresponding coordinate surfaces, one particular surface is chosen to which the value of $q_3 = 0$ is assigned. For instance in ellipsoidal coordinates the generating ellipsoid $\mathcal{E}$ is the height datum for ellipsoidal heights. Most of the other coordinate systems however try to use as reference for the height coordinate one particular equipotential surface of the gravity field. This is traditionally called the geoid and its choice, in the family of equipotentials, will necessarily occupy us in the book. In fact such a choice is not univocal and a lot of confusion has been generated by the practical custom of different nations to choose their own particular reference surface.

Nowadays with the important improvement in the knowledge of the global gravity field coming from space missions like GRACE (Tapley and Reigber 2001) and GOCE (Drinkwater et al. 2003), the time has come to make a precise choice defining one common height datum for the whole planet (Ihde et al. 2017).

While preparing the final issue of the book the authors have become aware of the existence of the work by Eremeev and Yurkina (1974). We would like to underline the notable closeness of that work to the actual spirit of this book, although the tools employed today take advantage of 50 years of geodetic research.

References

Anderson D.L. (2007). *New theory of the Earth*. Cambridge University Press, New York.
Bomford G. (1952). *Geodesy*. Clarendon Press, Oxford.
Drinkwater M.R., Floberghagen R., Haagmans R., Muzi D., Popescu A. (2003). *GOCE: ESA's first Earth Explorer Core mission*. In: Beutler G., Drinkwater M.R., Rummel R., Von Steiger R. (eds), *Earth gravity field from space-from sensors to Earth science*, Space Sciences Series of ISSI, 17:419–432, Springer, Dordrecht.
Eremeev V.F., Yurkina M.I. (1974). *Theorie der Höhen im Gravitationsfeld der Erde*. Arbeiten aus dem Vermessungs und Kartenwesen der Deutschen Demokratischen Republik, Band 32, Teil 1. Geodätischer Dienst, Leipzig (in German).
Grafarend E.W. (1975). *Cartan frames and a foundation of physical geodesy*. In: Brosowski B., Martensen E. (eds), *Methoden und Verfahren der Mathematischen Physic*, 12:179–208, BI-Verlag, Mannheim.
Heiskanen W.A. and Moritz H. (1967). *Physical geodesy*. Freeman, San Francisco.
Hotine M. (1969). *Mathematical geodesy*. ESSA Monograph 2, U.S. Department of Commerce, Washington, DC.
Ihde J., Sànchez L., Barzaghi R., Drewes H., Förste C., Gruber T., Liebsch G., Marti U., Pail R., Sideris M. (2017). *Definition and proposed realization of the International Height Reference System (IHRS)*. Surveys in Geophysics, 38(3):549–570.
Krarup T. (2006). *Letters on Molodenskiy's problem, III*. In: Borre K. (Ed.) Mathematical Foundation of Geodesy. Springer Verlag, Berlin, Heidelberg.
Lambeck K. (1988). *Geophysical geodesy: The slow deformations of the Earth*. Clarendon Press, Oxford.
Marussi A. (1985). *Intrinsic geodesy*. Springer, Berlin.
Sansò F., Sideris M.G. (2013). *Geoid determination: Theory and methods*. Lecture Notes in Earth System Sciences, Vol. 110. Springer-Verlag, Berlin, Heidelberg.
Tapley B., Reigber C. (2001). *The GRACE mission: status and future plans*. EOS. Trans. AGU,82(47), Fall Meet. Suppl., Abstract G41C-02.

Chapter 2
General Coordinates in $\mathcal{R}^3$

2.1 Outline

In this chapter we recall the basic definitions of coordinate systems in $\mathcal{R}^3$ and the related geometric concepts.

For the sake of completeness let us specify that when we talk about $\mathcal{R}^3$, we mean the classical Cartesian space where vectors are arrows, the modulus is the length of the arrow and the scalar product between two vectors is the usual product of their modulus times the cosine of the angle between the two.

Among other things, we have then that the shift between two points P and $\mathrm{P}' = \mathrm{P} + \delta\mathrm{P}$ can be identified with a vector $\delta\boldsymbol{r} = \delta\mathrm{P}$ and we know how to compute its modulus, $|\delta\boldsymbol{r}|$. This will be used in the chapter.

The chapter then focuses on orthogonal coordinate systems and the most basic examples of coordinates used in Geodesy, namely Cartesian, spherical and ellipsoidal, are presented, studied as well as the transformations between them.

2.2 Definitions and Reminds

A coordinate system on (a subset of) $\mathcal{R}^3$ is a triple of functions of the point P $\{q_1(\mathrm{P}), q_2(\mathrm{P}), q_3(\mathrm{P})\}$ that we shall arrange in an algebraic vector:

$$\boldsymbol{q}(\mathrm{P}) = \begin{vmatrix} q_1(\mathrm{P}) \\ q_2(\mathrm{P}) \\ q_3(\mathrm{P}) \end{vmatrix} , \tag{2.1}$$

such that the correspondence between $\boldsymbol{q}$ and P is biunivocal on (a subset of) $\mathcal{R}^3$, apart maybe from an exceptional subset of points which is explicitly excluded.

The above basic property of $\boldsymbol{q}(\mathrm{P})$ is typically complemented with some regularity hypothesis, that usually is that the functions $q_i(\mathrm{P})$ have a continuous gradient, namely

F. Sansò et al., *Geodetic Heights*, Springer Geophysics,
https://doi.org/10.1007/978-3-030-10454-2_2

that $\boldsymbol{q}(\mathrm{P})$ is a diffeomorphism. For the sake of completeness, we remind that $\nabla q_i(\mathrm{P})$ is by definition an $\mathcal{R}^3$ vector, when it exists, such that

$$\delta q_i(\mathrm{P}) = q_i(\mathrm{P} + \delta P) - q_i(\mathrm{P}) = \nabla q_i(\mathrm{P}) \cdot \delta \boldsymbol{r} + o(|\delta \boldsymbol{r}|) \tag{2.2}$$

an expression this which is meaningful thanks to the remark done in Sect. 2.1. We shall make the hypothesis that for every P in $\mathcal{R}^3$, or in the subset of our interest,

$$\mathrm{Span}\{\nabla q_i(\mathrm{P}), i = 1, 2, 3\} \equiv \mathcal{R}^3 \tag{2.3}$$

i.e. $\{\nabla q_i(\mathrm{P})\}$ is a basis in $\mathcal{R}^3$ [(1)].

This implies that $|\nabla q_i| \neq 0$ for $\forall \mathrm{P}$, $\forall i$; furthermore if we define the Jacobian operator $\mathcal{R}^3 \to \mathcal{R}^3$ by

$$J(\mathrm{P})\mathrm{d}\boldsymbol{r} = \begin{vmatrix} \nabla q_1(\mathrm{P})^T \mathrm{d}\boldsymbol{r} \\ \nabla q_2(\mathrm{P})^T \mathrm{d}\boldsymbol{r} \\ \nabla q_3(\mathrm{P})^T \mathrm{d}\boldsymbol{r} \end{vmatrix} = \begin{vmatrix} \nabla q_1 \cdot \mathrm{d}\boldsymbol{r} \\ \nabla q_2 \cdot \mathrm{d}\boldsymbol{r} \\ \nabla q_3 \cdot \mathrm{d}\boldsymbol{r} \end{vmatrix} = \mathrm{d}\boldsymbol{q}(\mathrm{P})\,[\mathrm{d}\boldsymbol{r}] \tag{2.4}$$

we see that J has to be invertible, $\forall \mathrm{P}$, since

$$J\mathrm{d}\boldsymbol{r} = 0 \Rightarrow \nabla q_i \cdot \mathrm{d}\boldsymbol{r} = 0 \Rightarrow \mathrm{d}\boldsymbol{r} = 0\,.$$

Given the system $\{q_i\}$ we can define three coordinate surfaces $\{\mathrm{S}_i\}$ as

$$\mathrm{S}_i = \{\mathrm{P};\ q_i(\mathrm{P}) = \mathrm{const}\}\,. \tag{2.5}$$

Through every point P_0 pass three coordinate surfaces $\mathrm{S}_i = \{\mathrm{P}; q_i(\mathrm{P}) = q_i(\mathrm{P}_0)\}$ and we see that $\nabla q_i(\mathrm{P}_0)$ are orthogonal to such surfaces. In fact take for instance S_1; if $\boldsymbol{t}$ is a unit vector tangent to S_1 at P_0, then

$$\boldsymbol{t} = \lim_{\varepsilon \to 0} \frac{\delta \boldsymbol{r}}{\varepsilon} \quad (\varepsilon = |\delta \boldsymbol{r}|)$$

where $\mathrm{P}_0 + \delta\mathrm{P} = \mathrm{P}_0 + \delta\boldsymbol{r} \in \mathrm{S}_1$. But then

$$\nabla q_1(\mathrm{P}_0) \cdot \boldsymbol{t} = \lim_{\varepsilon \to 0} \frac{1}{\varepsilon}\,[\nabla q_1(\mathrm{P}_0) \cdot \delta \boldsymbol{r}] = \lim_{\varepsilon \to 0} \frac{1}{\varepsilon}\,[\delta q_1 - o(\varepsilon)] = -\lim_{\varepsilon \to 0} \frac{o(\varepsilon)}{\varepsilon} = 0$$

because $\delta q_1 = q_1(\mathrm{P}_0 + \delta\mathrm{P}) - q_1(\mathrm{P}_0) = 0$, by (2.5). Similarly it happens for S_2, S_3.

[1] *Note*: Let us remind that in a linear vector space X, the $\mathrm{Span}\{\boldsymbol{x}_i,\ i = 1, 2, \ldots, N\}$ is just the linear subspace generated by the linear combinations $\{\sum_{i=1}^{N} \lambda_i \boldsymbol{x}_i,\ \lambda_i \in \mathcal{R},\ \forall i\}$; when the space becomes infinite dimensional, the $\mathrm{Span}\{\boldsymbol{x}_i,\ i = 1, 2, \ldots\}$ is just the subspace of all such finite dimensional linear combinations.

The intersection of two coordinate surfaces is a line, called coordinate line. Let us consider

$$\ell_{q_1} \equiv \ell_1 = \{S_2 \cap S_3\} ;$$

since $\ell_1 \subset S_2$, q_2 does not change along ℓ_1 and the same happens for q_3. So ℓ_1 is the line along which only q_1 varies, while q_2, q_3 remain constant; for this reason we have also denoted ℓ_1 as ℓ_{q_1}.

The same holds for all three combinations of indexes (note that $S_k \cap S_j \equiv S_j \cap S_k$). Now let $P_0 \equiv \{q_{01}, q_{02}, q_{03}\} \in \ell_1$; then the parametric form of ℓ_1 is

$$\ell_1 \equiv \{\boldsymbol{r}\,(q_1, q_{02}, q_{03})\} .$$

Passing to the first order differential formula one has

$$\mathrm{d}_1\boldsymbol{r} = \boldsymbol{r}\,(q_{01} + \mathrm{d}q_1, q_{02}, q_{03}) - \boldsymbol{r}\,(q_{01}, q_{02}, q_{03}) = \frac{\partial \boldsymbol{r}}{\partial q_1}\,(P_0)\,\mathrm{d}q_1 .$$

This generalizes to

$$\mathrm{d}_i\boldsymbol{r} = \frac{\partial \boldsymbol{r}}{\partial q_i}\,(P_0)\,\mathrm{d}q_i , \tag{2.6}$$

which shows that the three vectors

$$\boldsymbol{g}_i\,(P_0) = \frac{\partial \boldsymbol{r}}{\partial q_i}\,(P_0) \qquad i = 1, 2, 3 \tag{2.7}$$

are tangent to the three coordinate lines.

Moreover one has for any $\mathrm{d}\boldsymbol{r}$

$$\mathrm{d}\boldsymbol{r} = \sum_{i=1}^{3} \mathrm{d}_i\boldsymbol{r} = \sum_{i=1}^{3} \boldsymbol{g}_i \mathrm{d}q_i , \tag{2.8}$$

which is nothing but the theorem of the total differential.

If we arrange $\{\boldsymbol{g}_i\}$ in a matrix

$$\mathcal{H} = \left[\boldsymbol{g}_1, \boldsymbol{g}_2, \boldsymbol{g}_3\right] , \tag{2.9}$$

which is also a linear operator $\mathcal{R}^3 \to \mathcal{R}^3$ such that

$$\mathrm{d}\boldsymbol{r} = \mathcal{H} \begin{vmatrix} \mathrm{d}q_1 \\ \mathrm{d}q_2 \\ \mathrm{d}q_3 \end{vmatrix} = \mathcal{H}\mathrm{d}\boldsymbol{q} , \tag{2.10}$$

we can draw a very interesting conclusion. In fact, recalling (2.4), one can write

$$\mathrm{d}\boldsymbol{r} = \mathcal{H}\mathrm{d}\boldsymbol{q} = \mathcal{H}J\mathrm{d}\boldsymbol{r} \ , \tag{2.11}$$

and such a relation holds $\forall \mathrm{d}\boldsymbol{r}$.

Therefore one has

$$\mathcal{H}J = I \ , \tag{2.12}$$

namely

$$\mathcal{H} = J^{-1} \ , \tag{2.13}$$

which we know to exist thanks to a previous remark. The conclusion is at once that also $\mathcal{H}$ is a regular operator with a regular inverse. Hence we see that

$$g_i = \left|\boldsymbol{g}_i\right| \neq 0 \tag{2.14}$$

and $\{\boldsymbol{g}_i\,(\mathrm{P}_0)\,;\ i = 1, 2, 3\}$ is a basis of $\mathcal{R}^3$, $\forall \mathrm{P}_0$. This means that the three $\boldsymbol{g}_i\,(\mathrm{P}_0)$ are always linearly independent, i.e. that ℓ_i is never tangent to S_i. Note that this allows to define the *Cartan frame* of unit vectors $\{\boldsymbol{e}_1, \boldsymbol{e}_2, \boldsymbol{e}_3\}$ (Grafarend 1975), that are respectively tangent to ℓ_i, by

$$\boldsymbol{e}_i = \frac{\boldsymbol{g}_i}{g_i} \ . \tag{2.15}$$

Another remarkable consequence of (2.12) is that, since $\forall k$

$$\mathcal{H}J\boldsymbol{g}_k = \sum_{i=1}^{3} \boldsymbol{g}_i \nabla q_i^T \boldsymbol{g}_k = \sum_{i=1}^{3} \boldsymbol{g}_i \left(\nabla q_i \cdot \boldsymbol{g}_k\right) \equiv \boldsymbol{g}_k \ ,$$

we must also have

$$\nabla q_i \cdot \boldsymbol{g}_k = \delta_{ik} \ , \tag{2.16}$$

namely $\{\nabla q_i\}$, $\{\boldsymbol{g}_i\}$ are complementary bases or, in other words, they form together a biorthogonal system.

Some of the above entities are represented in Fig. 2.1.

Another fundamental geometric entity that descends from (2.10) is the metric tensor, which is intimately related to the modulus of $\mathrm{d}\boldsymbol{r}$. In fact we have

$$|\mathrm{d}\boldsymbol{r}|^2 = \mathrm{d}\boldsymbol{q}^T \mathcal{H}^T \mathcal{H} \mathrm{d}\boldsymbol{q} = \sum_{i,k=1}^{3} \mathrm{d}q_i \,\mathrm{d}q_k \,\boldsymbol{g}_i \cdot \boldsymbol{g}_k \ ; \tag{2.17}$$

the matrix G

$$G \equiv \left\{\boldsymbol{g}_i \cdot \boldsymbol{g}_k\right\} = \mathcal{H}^T \mathcal{H} \tag{2.18}$$

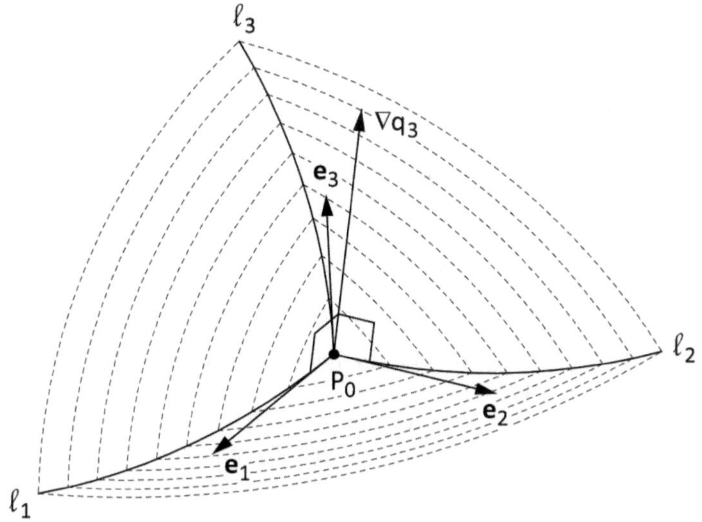

Fig. 2.1 The coordinate lines ℓ_i and the coordinate surfaces S_i at P_0; the Cartan frame $\{\boldsymbol{e}_1, \boldsymbol{e}_2, \boldsymbol{e}_3\}$ and $\nabla q_3 \perp S_3$

is by definition the metric tensor in $\mathcal{R}^3$ with respect to the coordinate system (2.18). As it is obvious from the definition (2.18), G is a symmetric positive definite matrix. Among other things, the metric tensor supplies the relation between the two bases $\{\boldsymbol{g}_i\}$, $\{\nabla q_i\}$.

In fact (2.12), also recalling (2.4) and (2.10), can be rewritten as

$$\mathcal{H}J = \sum_{i=1}^{3} \boldsymbol{g}_i \nabla q_i^T = I = J^T \mathcal{H}^T = \sum_{i=1}^{3} \nabla q_i \, \boldsymbol{g}_i^T , \tag{2.19}$$

giving

$$\boldsymbol{g}_k = \sum_{i=1}^{3} \nabla q_i \, \boldsymbol{g}_i^T \boldsymbol{g}_k = \sum_{i=1}^{3} \nabla q_i \, G_{ik} \tag{2.20}$$

or

$$\nabla q_i = \sum_{i=1}^{3} G_{ik}^{(-1)} \boldsymbol{g}_k \, . \tag{2.21}$$

Notice that, since $\mathcal{H}$ is invertible, G has to be invertible too; whence the correctness of (2.21).

Fortunately in Geodesy we mainly need, as purely geometric coordinates, only systems which are orthogonal, thus simplifying many useful formulas. As a definition, a system of coordinates is orthogonal when the three coordinate lines, meeting at any point P, cross orthogonally each other. This implies that the vectors $\boldsymbol{g}_i$, which are tangent to ℓ_i, are also orthogonal to one another. As a consequence the metric tensor becomes diagonal, namely

$$G_{ik} = \boldsymbol{g}_i \cdot \boldsymbol{g}_k = \boldsymbol{g}_i^2 \delta_{ik} \,. \tag{2.22}$$

In this case the inverse matrix, G^{-1}, is diagonal too and (2.21) becomes

$$\nabla q_i = \sum_{k=1}^{3} g_i^{-2} \delta_{ik} \, \boldsymbol{g}_k = \frac{\boldsymbol{g}_i}{g_i^2} = \frac{\boldsymbol{e}_i}{g_i} \,. \tag{2.23}$$

So the basis $\{\nabla q_i\}$ is just parallel to $\{\boldsymbol{g}_i\}$ and the triad of the Cartan frame $\{\boldsymbol{e}_i\}$ is parallel to the two and orthonormal.

One of the useful results of this relation is that the differential operator, ∇, has a quite simple representation in this case. In fact, by applying the chain rule, we find that $\forall F$ continuously differentiable

$$\nabla F = \sum_{i=1}^{3} \frac{\partial F}{\partial q_i} \nabla q_i = \sum_{i=1}^{3} \frac{\boldsymbol{e}_i}{g_i} \frac{\partial F}{\partial q_i} \,;$$

this implies that

$$\nabla = \sum_{i=1}^{3} \frac{\boldsymbol{e}_i}{g_i} \frac{\partial}{\partial q_i} \,. \tag{2.24}$$

Also for later reference, we note here that in curvilinear coordinates the expression of the Laplace operator,

$$\Delta = \nabla \cdot \nabla \,,$$

becomes then

$$\Delta = \frac{1}{G} \sum_{j=1}^{3} \frac{\partial}{\partial q_j} \left[\frac{G}{g_j^2} \frac{\partial}{\partial q_j} \right] \qquad (G = g_1 \, g_2 \, g_3) \,. \tag{2.25}$$

A proof can be found in Sansò and Sideris (2013, Chap.1, A3).

2.3 Relevant Examples

In this section we apply the general concepts presented in Sect. 2.2 to three families of $\mathcal{R}^3$ coordinates, namely Cartesian, spherical and ellipsoidal since all of them are useful, and used, in Geodesy.

2.3.1 Cartesian Coordinates

They are characterized by a triad of unit vectors, $(\boldsymbol{e}_x, \boldsymbol{e}_y, \boldsymbol{e}_z)$, orthogonal one to the other, issuing from a certain point O (origin) of the space $\mathcal{R}^3$ (see Fig. 2.2).

Given any point P in $\mathcal{R}^3$, we obtain the three coordinates (x, y, z) by orthogonal projection of P on the three axis, so obtaining P_x, P_y, P_z in Fig. 2.2. Then the linear coordinates of these three points on the three axes are exactly our coordinates.

It is immediate to verify that

$$x = \boldsymbol{r} \cdot \boldsymbol{e}_x\ , \quad y = \boldsymbol{r} \cdot \boldsymbol{e}_y\ , \quad z = \boldsymbol{r} \cdot \boldsymbol{e}_z \tag{2.26}$$

and then

$$\boldsymbol{r} = x\boldsymbol{e}_x + y\boldsymbol{e}_y + z\boldsymbol{e}_z\ . \tag{2.27}$$

Note that the notions of orthogonality and scalar product can be safely used on account of a remark at the beginning of Sect. 2.1.

Differentiating (2.27) and considering that $\boldsymbol{e}_x, \boldsymbol{e}_y, \boldsymbol{e}_z$ are constant vectors, we get

$$\mathrm{d}\boldsymbol{r} = \mathrm{d}x\boldsymbol{e}_x + \mathrm{d}y\boldsymbol{e}_y + \mathrm{d}z\boldsymbol{e}_z\ , \tag{2.28}$$

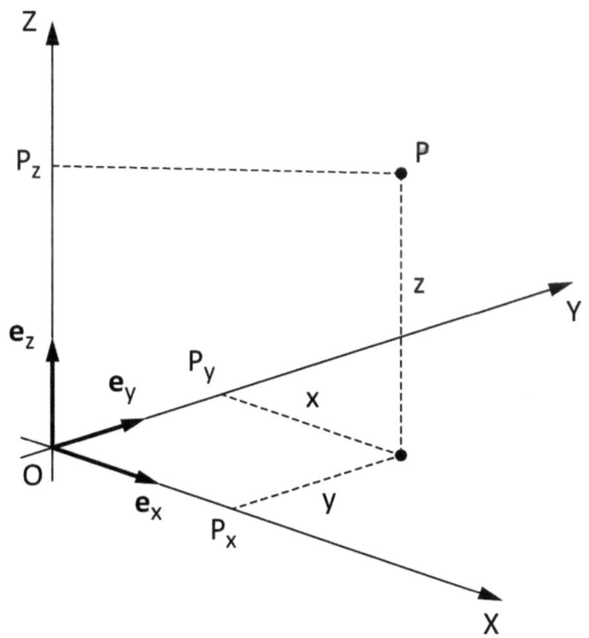

Fig. 2.2 The geometric construction of Cartesian coordinates

that compared with (2.8) gives at once

$$\boldsymbol{g}_1 = \boldsymbol{e}_1 = \boldsymbol{e}_x \ , \quad \boldsymbol{g}_2 = \boldsymbol{e}_2 = \boldsymbol{e}_y \ , \quad \boldsymbol{g}_3 = \boldsymbol{e}_3 = \boldsymbol{e}_z \ . \tag{2.29}$$

Accordingly we have too

$$g_i = 1 \ , \qquad G_{ik} = \boldsymbol{g}_i \cdot \boldsymbol{g}_k = \delta_{ik} \ ,$$

so that the ordinary Pythagorean relation is reconstructed

$$\mathrm{d}r^2 = \mathrm{d}x^2 + \mathrm{d}y^2 + \mathrm{d}z^2 \ . \tag{2.30}$$

Therefore gradient and Laplacian operators in Cartesian coordinates take the form

$$\nabla = \boldsymbol{e}_x \frac{\partial}{\partial x} + \boldsymbol{e}_y \frac{\partial}{\partial y} + \boldsymbol{e}_z \frac{\partial}{\partial z} \ , \tag{2.31}$$

$$\Delta = \frac{\partial^2}{\partial x^2} + \frac{\partial^2}{\partial y^2} + \frac{\partial^2}{\partial z^2} \ . \tag{2.32}$$

Since the item is so much in use in Geodesy, we shall dwell a little on the problem of deriving the transformation from a Cartesian system to another. We assume that in both systems the same unit length is used. Looking at Fig. 2.3 we see first that the vector relation holds

$$\boldsymbol{r} = \boldsymbol{t} + \boldsymbol{s} \ . \tag{2.33}$$

We write (2.33) in terms of components as

$$x\boldsymbol{e}_x + y\boldsymbol{e}_y + z\boldsymbol{e}_z = t_x\boldsymbol{e}_x + t_y\boldsymbol{e}_y + t_z\boldsymbol{e}_z + \xi\boldsymbol{e}_\xi + \eta\boldsymbol{e}_\eta + \zeta\boldsymbol{e}_\zeta \ . \tag{2.34}$$

Taking the scalar product of (2.34) with $\boldsymbol{e}_x$, $\boldsymbol{e}_y$, $\boldsymbol{e}_z$ respectively, we obtain the matrix relation

$$\begin{vmatrix} x \\ y \\ z \end{vmatrix} = \begin{vmatrix} t_x \\ t_y \\ t_z \end{vmatrix} + \begin{bmatrix} \boldsymbol{e}_x \cdot \boldsymbol{e}_\xi & \boldsymbol{e}_x \cdot \boldsymbol{e}_\eta & \boldsymbol{e}_x \cdot \boldsymbol{e}_\zeta \\ \boldsymbol{e}_y \cdot \boldsymbol{e}_\xi & \boldsymbol{e}_y \cdot \boldsymbol{e}_\eta & \boldsymbol{e}_y \cdot \boldsymbol{e}_\zeta \\ \boldsymbol{e}_z \cdot \boldsymbol{e}_\xi & \boldsymbol{e}_z \cdot \boldsymbol{e}_\eta & \boldsymbol{e}_z \cdot \boldsymbol{e}_\zeta \end{bmatrix} \begin{vmatrix} \xi \\ \eta \\ \zeta \end{vmatrix} \ . \tag{2.35}$$

It is clear that the vector

$$\begin{vmatrix} t_x \\ t_y \\ t_z \end{vmatrix}$$

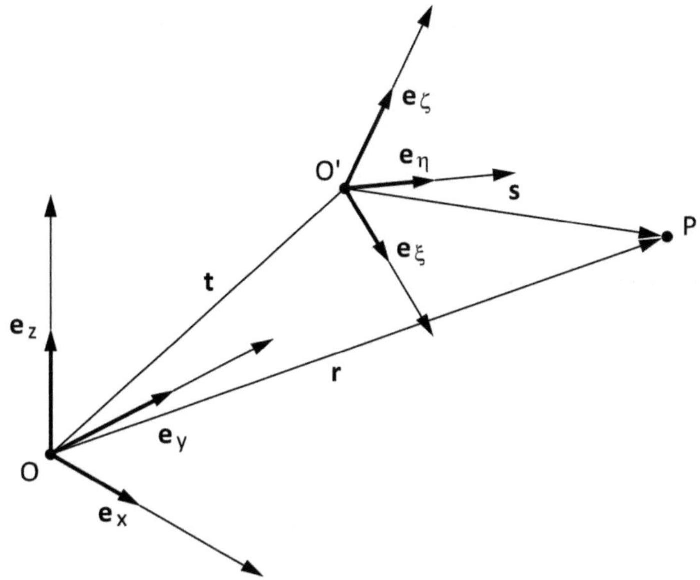

Fig. 2.3 The geometry of a roto-translation

represents the translation of the origin from O to O′, while the matrix

$$U = \begin{bmatrix} \boldsymbol{e}_x \cdot \boldsymbol{e}_\xi & \boldsymbol{e}_x \cdot \boldsymbol{e}_\eta & \boldsymbol{e}_x \cdot \boldsymbol{e}_\zeta \\ \boldsymbol{e}_y \cdot \boldsymbol{e}_\xi & \boldsymbol{e}_y \cdot \boldsymbol{e}_\eta & \boldsymbol{e}_y \cdot \boldsymbol{e}_\zeta \\ \boldsymbol{e}_z \cdot \boldsymbol{e}_\xi & \boldsymbol{e}_z \cdot \boldsymbol{e}_\eta & \boldsymbol{e}_z \cdot \boldsymbol{e}_\zeta \end{bmatrix}, \quad (2.36)$$

represents the rotation between the triad $(\boldsymbol{e}_\xi, \boldsymbol{e}_\eta, \boldsymbol{e}_\zeta)$ and the triad $(\boldsymbol{e}_x, \boldsymbol{e}_y, \boldsymbol{e}_z)$. By exchanging (ξ, η, ζ) with (x, y, z), we find the inverse rotation, i.e. the matrix

$$U^{-1} = \begin{bmatrix} \boldsymbol{e}_\xi \cdot \boldsymbol{e}_x & \boldsymbol{e}_\xi \cdot \boldsymbol{e}_y & \boldsymbol{e}_\xi \cdot \boldsymbol{e}_z \\ \boldsymbol{e}_\eta \cdot \boldsymbol{e}_x & \boldsymbol{e}_\eta \cdot \boldsymbol{e}_y & \boldsymbol{e}_\eta \cdot \boldsymbol{e}_z \\ \boldsymbol{e}_\zeta \cdot \boldsymbol{e}_x & \boldsymbol{e}_\zeta \cdot \boldsymbol{e}_y & \boldsymbol{e}_\zeta \cdot \boldsymbol{e}_z \end{bmatrix}. \quad (2.37)$$

We observe that the matrix (2.37) is also the transpose of U so that

$$U^T U \equiv U^{-1} U \equiv I , \quad (2.38)$$

namely U is an orthogonal matrix. This also reflects the fact that a rotation does not change the length of vectors; in fact

$$\left[U \begin{vmatrix} \xi \\ \eta \\ \zeta \end{vmatrix} \right]^T \left[U \begin{vmatrix} \xi \\ \eta \\ \zeta \end{vmatrix} \right] = [\, \xi \ \ \eta \ \ \zeta \,]\, U^T U \begin{vmatrix} \xi \\ \eta \\ \zeta \end{vmatrix} = \xi^2 + \eta^2 + \zeta^2 . \quad (2.39)$$

As it is well known, a rotation in $\mathcal{R}^3$ can always be defined by three angles, however instead of finding a general form of U, we prefer here to report it for the case that the rotation is infinitesimal. In such a case, in fact, we know that U should be close to the identity matrix, i.e. we can put

$$U = I + K \ , \tag{2.40}$$

with K an infinitesimal matrix. But then, by imposing (2.38) and neglecting the second order term $K^T K$, we get

$$K + K^T = 0 \ , \tag{2.41}$$

namely K is an infinitesimal anti-symmetric matrix

$$K = \begin{bmatrix} 0 & \alpha & \beta \\ -\alpha & 0 & \gamma \\ -\beta & -\gamma & 0 \end{bmatrix} . \tag{2.42}$$

So the infinitesimal rotation can be written as

$$U \begin{vmatrix} \xi \\ \eta \\ \zeta \end{vmatrix} = \begin{vmatrix} \xi \\ \eta \\ \zeta \end{vmatrix} + K \begin{vmatrix} \xi \\ \eta \\ \zeta \end{vmatrix} . \tag{2.43}$$

A final remark is that, by comparison of the components, one realizes that the components of

$$K \begin{vmatrix} \xi \\ \eta \\ \zeta \end{vmatrix}$$

are the same as those of the vector $\boldsymbol{\omega} \wedge \boldsymbol{s}$, where

$$\begin{cases} \boldsymbol{s} &= \xi \boldsymbol{e}_x + \eta \boldsymbol{e}_y + \zeta \boldsymbol{e}_z \\ \boldsymbol{\omega} &= -\gamma \boldsymbol{e}_x + \beta \boldsymbol{e}_y + -\alpha \boldsymbol{e}_z \end{cases} . \tag{2.44}$$

2.3.2 *Spherical Coordinates*

Instead of building spherical coordinates autonomously, we prefer to pass through a Cartesian system so that we shall define together the Cartesian system $(\boldsymbol{e}_x, \boldsymbol{e}_y, \boldsymbol{e}_z)$ and the attached spherical coordinate system.

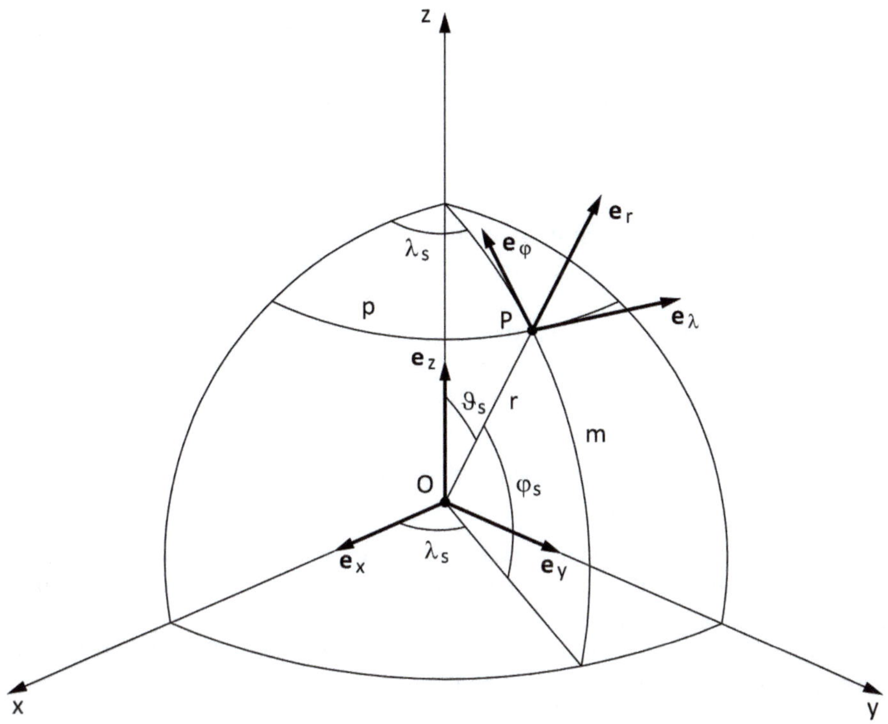

Fig. 2.4 The geometric of a spherical coordinate system; $(\boldsymbol{e}_\lambda, \boldsymbol{e}_\varphi, \boldsymbol{e}_r)$ Cartan frame at P

So we start from $(\boldsymbol{e}_x, \boldsymbol{e}_y, \boldsymbol{e}_z)$ and, given any point P, we first define the coordinate r, which is the distance of P from the origin O. Then we consider the angle between the position vector $\boldsymbol{r}$ and $\boldsymbol{e}_z$ and we call it the spherical colatitude, ϑ_s. Associated to this angle is the definition of spherical latitude φ_s (see Fig. 2.4),

$$\varphi_s = \frac{\pi}{2} - \vartheta_s \, . \tag{2.45}$$

Finally we consider the plane containing $\boldsymbol{r}$ and $\boldsymbol{e}_z$ and the plane containing $\boldsymbol{e}_x$ and $\boldsymbol{e}_z$; the dyhedral angle between the two is the spherical longitude λ_s (see Fig. 2.4).

The triple $(\lambda_s, \varphi_s, r)$ is the system of spherical coordinates of the point P. The plane of $\boldsymbol{e}_x$, $\boldsymbol{e}_y$ is called the equatorial plane, the axis z the polar axis.

Looking at Fig. 2.4, one realizes that the line ℓ_λ, with φ_s = const, r = const, is the circle p, called parallel, through P with centre on the z axis and contained in a plane parallel to the equatorial plane. The longitude λ_s has conventionally the range

$$0 \le \lambda_s < 360^\circ \text{ (or } 2\pi) \, . \tag{2.46}$$

The line ℓ_φ, with $\lambda_s = \text{const}$, $r = \text{const}$, called meridian, is half a circle m through P with center O, contained in the plane of $\boldsymbol{r}$ and $\boldsymbol{e}_z$, called meridian plane. The latitude has conventionally the range

$$-90° \le \varphi_s \le 90° \quad \left(-\frac{\pi}{2} \le \varphi_s \le \frac{\pi}{2}\right) . \tag{2.47}$$

The line ℓ_r, with $\lambda_s = \text{const}$, $\varphi_s = \text{const}$, is the half line that joins the origin O with P. The radial distance r has conventionally the range

$$0 \le r < +\infty . \tag{2.48}$$

The coordinate surfaces are respectively:

$S_\lambda =$ the meridian plane;
$S_\varphi =$ the cone with axis z projecting P from O (vertex);
$S_r =$ the sphere of radius r.

It is clear that this coordinate system is singular; in fact all the points of the z axis correspond to all the values of λ_s; moreover the origin, $r = 0$, corresponds to all the possible values of λ_s and φ_s.

A close look at Fig. 2.4 yields, by elementary geometry, the relation between spherical and Cartesian coordinates, namely

$$\boldsymbol{r} \sim \begin{vmatrix} x \\ y \\ z \end{vmatrix} = r \begin{vmatrix} \cos\varphi_s \cos\lambda_s \\ \cos\varphi_s \sin\lambda_s \\ \sin\varphi_s \end{vmatrix} . \tag{2.49}$$

With the symbol $\sim$ we mean the components of the vector in the Cartesian system. Such a relation is easily inverted by

$$\begin{cases} \lambda_s = \operatorname{atan}(y, x) \\ \varphi_s = \operatorname{arctg}\dfrac{z}{\sqrt{x^2+y^2}} \\ r \;\; = \sqrt{x^2+y^2+z^2} \end{cases} , \tag{2.50}$$

where by $\operatorname{atan}(y, x)$ we mean

$$\operatorname{atan}(y, x) = \begin{cases} \operatorname{arctg}\dfrac{y}{x} & x \ge 0 \\ \pi + \operatorname{arctg}\dfrac{y}{x} & x < 0 \end{cases} . \tag{2.51}$$

Note that, since $\operatorname{arctg} t$ ranges over $\left(-\frac{\pi}{2}, \frac{\pi}{2}\right)$, the second of (2.50) provides the correct values for the spherical latitude.

From (2.49) we can compute

$$\boldsymbol{g}_\lambda = \frac{\partial \boldsymbol{r}}{\partial \lambda} = \frac{\partial x}{\partial \lambda}\boldsymbol{e}_x + \frac{\partial y}{\partial \lambda}\boldsymbol{e}_y + \frac{\partial z}{\partial \lambda}\boldsymbol{e}_z \sim r\cos\varphi_s \left|\begin{array}{c} -\sin\lambda_s \\ \cos\lambda_s \\ 0 \end{array}\right| , \tag{2.52}$$

so that

$$g_\lambda = r\cos\varphi_s\ , \qquad \boldsymbol{e}_\lambda \sim \left|\begin{array}{c} -\sin\lambda_s \\ \cos\lambda_s \\ 0 \end{array}\right| . \tag{2.53}$$

Similarly

$$\boldsymbol{g}_\varphi \sim r \left|\begin{array}{c} -\sin\varphi_s\cos\lambda_s \\ -\sin\varphi_s\sin\lambda_s \\ \cos\varphi_s \end{array}\right| \tag{2.54}$$

$$g_\varphi = r\ , \qquad \boldsymbol{e}_\varphi \sim \left|\begin{array}{c} -\sin\varphi_s\cos\lambda_s \\ -\sin\varphi_s\sin\lambda_s \\ \cos\varphi_s \end{array}\right| \tag{2.55}$$

and

$$\boldsymbol{g}_r \sim \left|\begin{array}{c} \cos\varphi_s\cos\lambda_s \\ \cos\varphi_s\sin\lambda_s \\ \sin\varphi_s \end{array}\right| \tag{2.56}$$

$$g_r = 1\ , \qquad \boldsymbol{e}_r \sim \left|\begin{array}{c} \cos\varphi_s\cos\lambda_s \\ \cos\varphi_s\sin\lambda_s \\ \sin\varphi_s \end{array}\right| \tag{2.57}$$

It is immediate to verify that

$$\boldsymbol{e}_\lambda \cdot \boldsymbol{e}_\varphi = \boldsymbol{e}_\lambda \cdot \boldsymbol{e}_r = \boldsymbol{e}_\varphi \cdot \boldsymbol{e}_r = 0 \tag{2.58}$$

so that the spherical coordinate system turns out to be orthogonal.

From the above relations we then derive the shape of the metric, namely

$$\mathrm{d}s^2 = r^2\cos^2\varphi\,\mathrm{d}\lambda^2 + r^2\,\mathrm{d}\varphi^2 + \mathrm{d}r^2\ . \tag{2.59}$$

The shape of the gradient operator in spherical coordinates is a direct consequence of the above and of (2.24), namely

$$\nabla = \frac{1}{r\cos\varphi}\,\boldsymbol{e}_\lambda\,\frac{\partial}{\partial\lambda} + \frac{1}{r}\,\boldsymbol{e}_\varphi\,\frac{\partial}{\partial\varphi} + \boldsymbol{e}_r\,\frac{\partial}{\partial r}\ . \tag{2.60}$$

The shape of Laplacian is also easily derived from (2.25), although it could be usefully computed, as an exercise, by using the following differentiation table

$$\begin{cases} \dfrac{\partial \boldsymbol{e}_\lambda}{\partial \lambda} = -\cos\varphi\, \boldsymbol{e}_r + \sin\varphi\, \boldsymbol{e}_\varphi \\ \dfrac{\partial \boldsymbol{e}_\lambda}{\partial \varphi} = 0 \\ \dfrac{\partial \boldsymbol{e}_\lambda}{\partial r} = 0 \end{cases}, \tag{2.61}$$

$$\begin{cases} \dfrac{\partial \boldsymbol{e}_\varphi}{\partial \lambda} = -\sin\varphi\, \boldsymbol{e}_\lambda \\ \dfrac{\partial \boldsymbol{e}_\varphi}{\partial \varphi} = -\boldsymbol{e}_r \\ \dfrac{\partial \boldsymbol{e}_\varphi}{\partial r} = 0 \end{cases}, \tag{2.62}$$

$$\begin{cases} \dfrac{\partial \boldsymbol{e}_r}{\partial \lambda} = \cos\varphi\, \boldsymbol{e}_\lambda \\ \dfrac{\partial \boldsymbol{e}_r}{\partial \varphi} = \boldsymbol{e}_\varphi \\ \dfrac{\partial \boldsymbol{e}_r}{\partial r} = 0 \end{cases}. \tag{2.63}$$

In any event the result is $\left(G = r^2\cos\varphi\right)$, so that

$$\begin{aligned} \Delta = {} & \frac{1}{r^2\cos\varphi_s}\frac{\partial}{\partial\lambda}\frac{r^2\cos\varphi_s}{r^2\cos^2\varphi_s}\frac{\partial}{\partial\lambda} + \\ & + \frac{1}{r^2\cos\varphi_s}\frac{\partial}{\partial r}\frac{r^2\cos\varphi_s}{r^2}\frac{\partial}{\partial\varphi} + \\ & + \frac{1}{r^2\cos\varphi_s}\frac{\partial}{\partial r}r^2\cos\varphi_s\frac{\partial}{\partial r} = \\ = {} & \frac{\partial^2}{\partial r^2} + \frac{2}{r}\frac{\partial}{\partial r} + \frac{1}{r^2}\left(\frac{\partial^2}{\partial\varphi^2} - \operatorname{tg}\varphi_s\frac{\partial}{\partial\varphi} + \frac{1}{r^2\cos^2\varphi_s}\frac{\partial^2}{\partial\lambda^2}\right). \end{aligned} \tag{2.64}$$

Concluding we recall that the angular part of (2.64), namely the term in brackets, is called in literature the Laplace-Beltrami operator

$$\Delta_\sigma = \frac{\partial^2}{\partial\varphi^2} - \operatorname{tg}\varphi_s\frac{\partial}{\partial\varphi} + \frac{1}{r^2\cos^2\varphi_s}\frac{\partial^2}{\partial\lambda^2}. \tag{2.65}$$

Indeed if we use the colatitude ϑ_s instead of the latitude φ_s, (2.60) and (2.64) become

$$\begin{cases} \nabla = \dfrac{1}{r \sin \vartheta_s} \, \boldsymbol{e}_\lambda \, \dfrac{\partial}{\partial \lambda} + \dfrac{1}{r} \, \boldsymbol{e}_\vartheta \, \dfrac{\partial}{\partial \vartheta} + \boldsymbol{e}_r \, \dfrac{\partial}{\partial r} \\ \Delta = \dfrac{\partial^2}{\partial r^2} + \dfrac{2}{r} \, \dfrac{\partial}{\partial r} + \dfrac{1}{r^2} \left(\dfrac{\partial^2}{\partial \vartheta^2} + \operatorname{cotg} \vartheta_s \, \dfrac{\partial}{\partial \vartheta} + \dfrac{1}{r^2 \sin^2 \vartheta_s} \, \dfrac{\partial^2}{\partial \lambda^2} \right) \end{cases} . \tag{2.66}$$

2.3.3 Ellipsoidal Coordinates

This type of coordinates is particularly important in Geodesy because there is an equipotential surface of the gravity field which is in a relative distance of the order of 10^{-5} from an oblate ellipsoid, the same on which we base our system. Yet, here we shall look at ellipsoidal coordinates from a purely geometric point of view.

Our starting point is an oblate ellipsoid $\mathcal{E}$ with a certain system attached to it. The z axis is along the minor axis of $\mathcal{E}$ and it is also its symmetry axis. The origin of the Cartesian system is placed at the centre of $\mathcal{E}$, so that the $\boldsymbol{e}_x$, $\boldsymbol{e}_y$ plane, called the equatorial plane, cuts $\mathcal{E}$ along a circle, the equator, with maximum radius equal to a, while the z axis cuts $\mathcal{E}$ at two points, at distance b from the origin, called respectively the north and the south pole. If we cut $\mathcal{E}$ with planes parallel to the equatorial plane, in a distance smaller or equal to b from it, we obtain circles, with the centre on the z axis, called parallels. If we cut $\mathcal{E}$ with planes containing the z axis, we obtain ellipses, equal to one another due to the cylindrical (rotational) symmetry around z; each half ellipse (Fig. 2.5) is called a meridian and its half plane, with border z, is a meridian plane.

Now consider the space outside $\mathcal{E}$ and a layer internal to $\mathcal{E}$ of points P at a distance from the surface smaller than the minimum curvature radius of $\mathcal{E}$; all these points P have a unique orthogonal projection P_e on $\mathcal{E}$, i.e. the segment P_eP is aligned with the normal $\boldsymbol{\nu}$ of $\mathcal{E}$ at P_e.

Another characterization of P_e is that it is the point on $\mathcal{E}$ at minimum distance from P. We shall identify later the points in space that do not satisfy this property.

Now we define the longitude λ of P as the dyhedral angle between the $(\boldsymbol{e}_x, \boldsymbol{e}_z)$ plane (this is fixed conventionally once and for all) and the half plane containing the z axis and P; we explicitly mention that such a plane, that we call meridian, contains as well the normal $\boldsymbol{\nu}$ to $\mathcal{E}$ at P_e. The longitude λ, as for the spherical case, spans the interval

$$0 \leq \lambda < 360^\circ \; (\equiv 2\pi) \, .$$

Now consider the normal to $\mathcal{E}$ passing through P; such a line belongs to the meridian plane and crosses the equatorial plane at some point P_0 (see Fig. 2.5). The angle between $\boldsymbol{\nu}$ and the equatorial plane is called latitude φ of P. Also the latitude, similarly to the spherical case, spans the range

$$\left(-\frac{\pi}{2} \equiv\right) -90^\circ \leq \varphi_s \leq 90^\circ \left(\equiv \frac{\pi}{2}\right) .$$

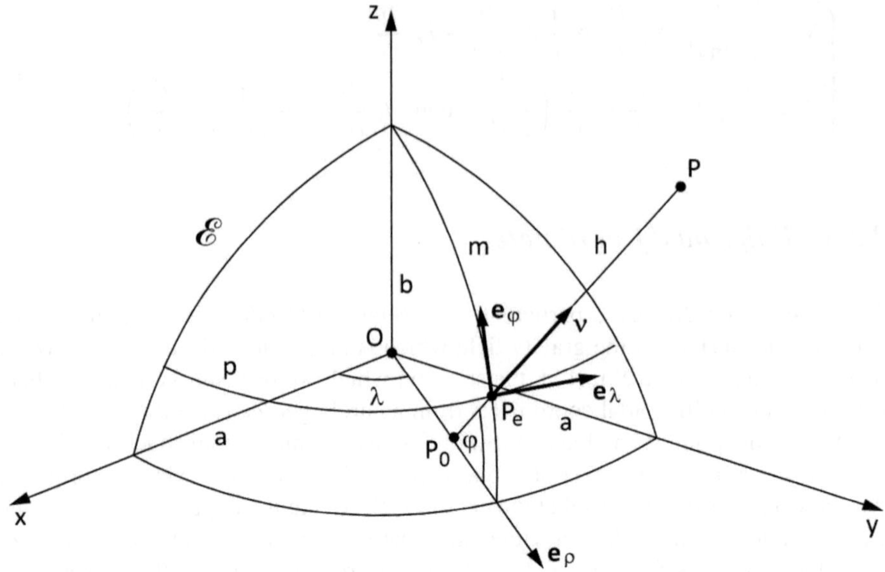

Fig. 2.5 The geometry of the reference ellisoid $\mathcal{E}$: a,b semimajor and semiminor axes, λ longitude, φ latitude, $h = \overline{\mathrm{P}_e\mathrm{P}}$ ellipsoidal height, p parallel, m meridian, $(\boldsymbol{e}_\lambda, \boldsymbol{e}_\varphi, \boldsymbol{\nu})$ Cartan frame on $\mathcal{E}$, $\boldsymbol{e}_\rho$ radial unit vector in the equatorial plane

Sometimes, we rather consider the colatitude ϑ of P defined as

$$\vartheta = \frac{\pi}{2} - \varphi ,$$

namely the angle between $\boldsymbol{\nu}$ and $\boldsymbol{e}_z$. Indeed ϑ sweeps the interval

$$0 \le \vartheta \le 180^\circ \; (\equiv \pi) ,$$

while φ runs from 90° to −90°, i.e. from the north to the south pole.

Finally we consider the distance $\overline{\mathrm{P}_e\mathrm{P}}$ and we define the third coordinate, the ellipsoidal height h, as such a distance for all P outside $\mathcal{E}$, or the distance changed of sign for points P internal to $\mathcal{E}$.

Our first purpose now is to find the transformation between the ellipsoidal triad (λ, φ, h) and (x, y, z) and its inverse. For this purpose we exploit the fact that $\mathcal{E}$ has a cylindrical symmetry. Then, after posing

$$x^2 + y^2 = \rho^2 ,$$

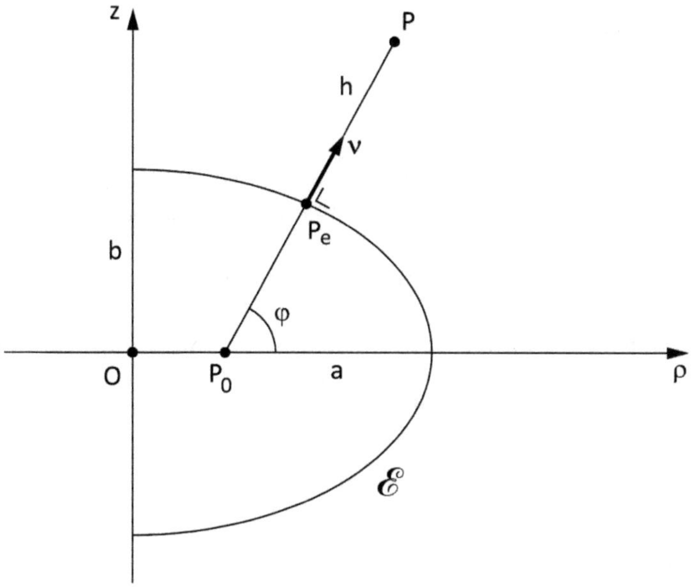

Fig. 2.6 The surface $\mathcal{E}$ cut by a meridian semiplane

we can study the couple of coordinates (ρ, z) in relation to (φ, h); once the transformation is written, we simply recall that

$$\begin{vmatrix} x \\ y \end{vmatrix} = \rho \begin{vmatrix} \cos\lambda \\ \sin\lambda \end{vmatrix} \tag{2.67}$$

and the 3D transformation is found, because

$$\boldsymbol{r} = z(\varphi, h)\,\boldsymbol{e}_z + \rho(\varphi, h)\,\boldsymbol{e}_\rho = z(\varphi, h)\,\boldsymbol{e}_z + \rho(\varphi, h)\left(\cos\lambda\boldsymbol{e}_x + \sin\lambda\boldsymbol{e}_y\right) .$$

So, inspecting Fig. 2.6, we see first that the ellipse $\mathcal{E}$ of the figure has equation

$$\frac{\rho^2}{a^2} + \frac{z^2}{b^2} = 1 . \tag{2.68}$$

We recall the definition of first eccentricity of $\mathcal{E}$, namely the parameter e derived from

$$e^2 = \frac{a^2 - b^2}{a^2} = 1 - \frac{b^2}{a^2} ; \tag{2.69}$$

from (2.69) we see as well that

$$\frac{b^2}{a^2} = 1 - e^2 .$$

So multiplying (2.68) by b^2, we find

$$\left(1-e^2\right)\,\rho^2+z^2=b^2\,. \tag{2.70}$$

Let the point P_e correspond to the latitude φ and assume we move it along the ellipse by an infinitesimal vector with components $(\mathrm{d}\rho,\,\mathrm{d}z)$ so that φ goes to $\varphi+\mathrm{d}\varphi$. Since we are on the ellipse, the differential of the left hand side of (2.70) has to be zero,

$$\left(1-e^2\right)\rho\,\mathrm{d}\rho+z\,\mathrm{d}z=0\,. \tag{2.71}$$

Indeed, calling $\boldsymbol{e}_\rho$, $\boldsymbol{e}_z$ the unit vectors along the two axes, $\mathrm{d}\rho\,\boldsymbol{e}_\rho+\mathrm{d}z\,\boldsymbol{e}_z$ is tangent to $\mathcal{E}$ in P_e; but then (2.71) shows that the vector $\left(1-e^2\right)\rho\,\boldsymbol{e}_\rho+z\,\boldsymbol{e}_z$ is orthogonal to the tangent, i.e. it is directed as the normal $\boldsymbol{\nu}$. This however implies that

$$\mathrm{tg}\varphi=\frac{z}{\left(1-e^2\right)\rho}\,. \tag{2.72}$$

Substituting $z=\left(1-e^2\right)\mathrm{tg}\varphi\,\rho$ in (2.70), we get

$$\rho^2=\frac{b^2}{\left(1-e^2\right)}\frac{1}{1-\left(1-e^2\right)\mathrm{tg}^2\varphi}=\frac{a^2\cos^2\varphi}{\cos^2\varphi+\left(1-e^2\right)\sin^2\varphi}\,;$$

simplifying and extracting the square root, yields

$$\rho=\frac{a\cos\varphi}{\sqrt{1-e^2\sin^2\varphi}}\,. \tag{2.73}$$

Putting (2.72), (2.73) together, we find

$$\begin{vmatrix}\rho\\ z\end{vmatrix}=\mathcal{N}\begin{vmatrix}\cos\varphi\\ \left(1-e^2\right)\sin\varphi\end{vmatrix} \tag{2.74}$$

$$\left(\mathcal{N}=\frac{a}{\sqrt{1-e^2\sin^2\varphi}}\right)\,.$$

The Eqs. (2.74) are the parametric equations of $\mathcal{E}$, expressed as functions of the parameter φ; so $(\rho,\,z)$ of this formula refer to the point P_e.

Noting that $\overline{\mathrm{P}_e\mathrm{P}}$ is $h\boldsymbol{\nu}$ and observing that $\boldsymbol{\nu}=\cos\varphi\,\boldsymbol{e}_\rho+\sin\varphi\,\boldsymbol{e}_z$ by definition of φ, we see that

$$\begin{vmatrix}\rho\\ z\end{vmatrix}_{\mathrm{P}}=\begin{vmatrix}(\mathcal{N}+h)\cos\varphi\\ \left[\left(1-e^2\right)\mathcal{N}+h\right]\sin\varphi\end{vmatrix}\,. \tag{2.75}$$

So exploiting (2.67), we finally get

$$\begin{vmatrix} x \\ y \\ z \end{vmatrix}_P = \begin{vmatrix} (\mathcal{N}+h)\cos\varphi\cos\lambda \\ (\mathcal{N}+h)\cos\varphi\sin\lambda \\ \left[\left(1-e^2\right)\mathcal{N}+h\right]\sin\varphi \end{vmatrix} , \tag{2.76}$$

which is the sought transformation, $(\lambda, \varphi, h) \to (x, y, z)$.

The inverse transformation is somewhat more intricate, however we have a closed algorithm providing it upon the definition of an intermediate angular variable ψ,

$$\begin{aligned} \mathrm{tg}\psi &= \frac{1}{\sqrt{1-e^2}}\frac{z}{\rho} \\ &\left(\rho^2 = x^2 + y^2\right) . \end{aligned} \tag{2.77}$$

With the help of (2.77) one finds

$$\begin{cases} \lambda = \mathrm{atan}\,(y, x) \\ \varphi = \mathrm{arctg}\dfrac{z + \left(e'\right)^2 b \sin^3\psi}{\rho - e^2 a \cos^3\psi} \\ h = \dfrac{\rho}{\cos\varphi} - \mathcal{N}(\varphi) \end{cases} ; \tag{2.78}$$

note that in the first of (2.78) we have used atan (,) defined by (2.51), while in the second equation we have a true arctg because φ is always between $-90°$ and $90°$; moreover we have employed here the second eccentricity e' defined by

$$\left(e'\right)^2 = \frac{a^2 - b^2}{b^2} . \tag{2.79}$$

A proof of (2.78) is cumbersome and not reported here.

We are now in a position to determine the two triads $\left(\boldsymbol{g}_\lambda, \boldsymbol{g}_\varphi, \boldsymbol{g}_h\right)$, $\left(\boldsymbol{e}_\lambda, \boldsymbol{e}_\varphi, \boldsymbol{e}_h\right)$ in terms of their Cartesian components. For this purpose we first prepare the differential formulas

$$\begin{aligned} &\frac{\partial}{\partial\varphi}\mathcal{N}(\varphi)\cos\varphi = -\mathcal{M}\sin\varphi , \\ &\left(1-e^2\right)\frac{\partial}{\partial\varphi}\mathcal{N}(\varphi)\sin\varphi = \mathcal{M}\cos\varphi , \\ &\mathcal{M} \equiv \frac{\left(1-e^2\right)a}{\left[1-e^2\sin^2\varphi\right]^{\frac{3}{2}}} \end{aligned} \tag{2.80}$$

and then we easily compute

$$
\begin{aligned}
\boldsymbol{g}_\lambda &= \frac{\partial \boldsymbol{r}}{\partial \lambda} \sim (\mathcal{N}+h)\cos\varphi \begin{vmatrix} -\sin\lambda \\ \cos\lambda \\ 0 \end{vmatrix} \\
g_\lambda &= (\mathcal{N}+h)\cos\varphi \qquad (2.81)\\
\boldsymbol{e}_\lambda &= \begin{vmatrix} -\sin\lambda \\ \cos\lambda \\ 0 \end{vmatrix} ;
\end{aligned}
$$

$$
\begin{aligned}
\boldsymbol{g}_\varphi &= \frac{\partial \boldsymbol{r}}{\partial \varphi} \sim (\mathcal{M}+h) \begin{vmatrix} -\sin\varphi\cos\lambda \\ -\sin\varphi\sin\lambda \\ \cos\varphi \end{vmatrix} \\
g_\varphi &= \mathcal{M}+h \qquad (2.82)\\
\boldsymbol{e}_\varphi &= \begin{vmatrix} -\sin\varphi\cos\lambda \\ -\sin\varphi\sin\lambda \\ \cos\varphi \end{vmatrix} ;
\end{aligned}
$$

$$
\begin{aligned}
\boldsymbol{g}_h &= \frac{\partial \boldsymbol{r}}{\partial h} \sim \begin{vmatrix} \cos\varphi\cos\lambda \\ \cos\varphi\sin\lambda \\ \sin\varphi \end{vmatrix} \\
g_h &= 1 \qquad (2.83)\\
\boldsymbol{e}_h &= \begin{vmatrix} \cos\varphi\cos\lambda \\ \cos\varphi\sin\lambda \\ \sin\varphi \end{vmatrix} .
\end{aligned}
$$

From the above formulas, it immediately descends that (λ, φ, h) is an orthogonal system of coordinates, because

$$\boldsymbol{e}_\lambda \cdot \boldsymbol{e}_\varphi = \boldsymbol{e}_\lambda \cdot \boldsymbol{e}_h = \boldsymbol{e}_\varphi \cdot \boldsymbol{e}_h = 0 .$$

Therefore the metric form in these coordinates is given by

$$\mathrm{d}s^2 = (\mathcal{N}+h)^2\cos^2\varphi\,\mathrm{d}\lambda^2 + (\mathcal{M}+h)^2\,\mathrm{d}\varphi^2 + \mathrm{d}h^2 \qquad (2.84)$$

and the gradient operator is

$$\nabla = \frac{\boldsymbol{e}_\lambda}{(\mathcal{N}+h)\cos\varphi}\frac{\partial}{\partial\lambda} + \frac{\boldsymbol{e}_\varphi}{(\mathcal{M}+h)}\frac{\partial}{\partial\varphi} + \boldsymbol{\nu}\frac{\partial}{\partial h} . \qquad (2.85)$$

The Laplacian, that can indeed be expressed in (λ, φ, h) coordinates, is however not useful because computations of the gravity field referred to the ellipsoid $\mathcal{E}$ are usually done in a different system of ellipsoidal coordinates, which is maybe less intuitive from the geometric point of view, but enjoys the notable property of separating the variables of the Laplacian itself (see Heiskanen and Moritz 1967, Sansò and Sideris 2013).

As for the geometric meaning of the two quantities $\mathcal{N}$, $\mathcal{M}$, let us observe that if we move P along its meridian (λ = const, h = const) with a latitude variation $\mathrm{d}\varphi$, from (2.80) and (2.84) we see that

$$\frac{\mathrm{d}s}{\mathrm{d}\varphi} = \sqrt{\frac{\mathrm{d}s^2}{\mathrm{d}\varphi^2}} = \sqrt{(\mathcal{M}+h)^2 \sin^2\varphi + (\mathcal{M}+h)^2 \cos^2\varphi} = \mathcal{M}+h\ , \tag{2.86}$$

implying that $\mathcal{M}+h$ is in fact the curvature radius of the meridian passing through P.

Similarly the second relation of (2.81), that gives the radius of the parallel p through P, tells us also that $\mathcal{N}+h$ is the curvature radius of the normal section (i.e. section of the surface h = const, with a plane containing $\boldsymbol{\nu}$) tangent to the parallel and orthogonal to the meridian.

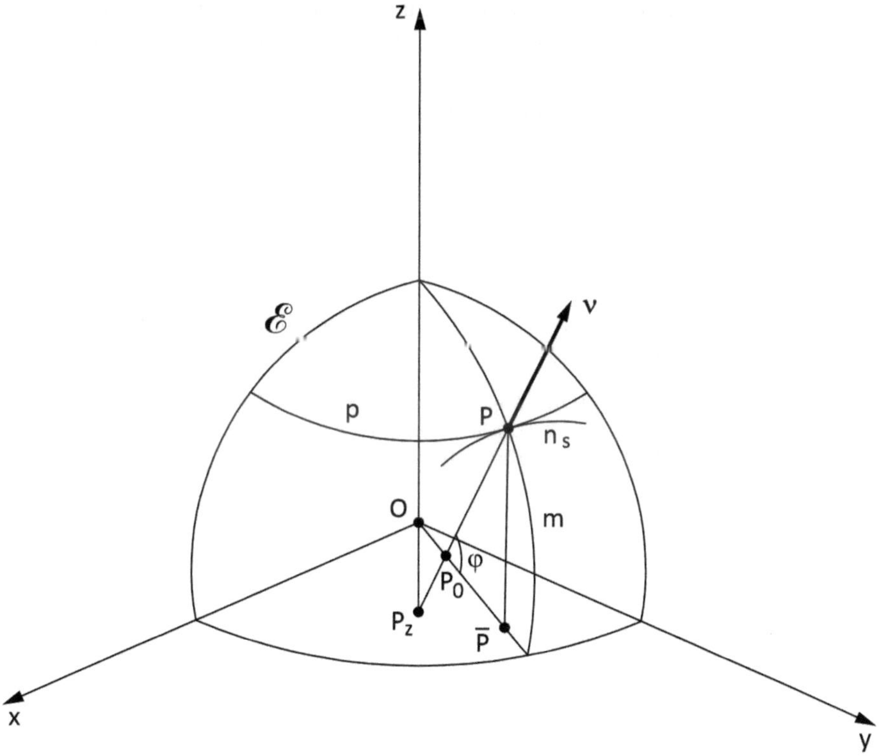

Fig. 2.7 $\mathcal{E}$ reference ellipsoid; $\boldsymbol{\nu}$ normal at P; $\overline{\text{OP}}$ radius of the parallel; n_s normal section tangent to p and orthogonal to m at P

The quantity $\mathcal{N}$ is also quoted in literature as the grand normal of the ellipsoid $\mathcal{E}$ at P_e and it represents as well the distance of P_e from the z axis along the normal $\boldsymbol{\nu}$, as one can see from (2.75), taking $\varphi = \text{const}$ and imposing $\rho = 0$.

Note that it holds $\mathcal{M} \leq \mathcal{N}$ for any φ.

The above statements are represented in Fig. 2.7.

Finally let us recall that our ellipsoidal coordinates have been defined by means of an orthogonal projection of point P on $\mathcal{E}$. Observe that by setting $z = 0$ in (2.75), one finds that P_0 (see Fig. 2.7) has a distance from the z axis $\rho_0 = e^2 \mathcal{N} \cos\varphi$; then one has for every $\varphi \neq 0$

$$\overline{P_0 P} = \mathcal{N}\left(1 - e^2\right) < \mathcal{M} < \mathcal{N} .$$

Therefore all the points of the segment P_0P are in a distance smaller than the two radius of curvature at P and therefore they project orthogonally on P. This one-to-one correspondence breaks down at P_0, which for symmetry reason is projected on both, P and its symmetric image on the southern hemiellipsoid. Since the maximum of ρ_0 is achieved at $\varphi = 0$ and it is equal to $\rho_0 = e^2 a$, we may conclude that all points internal to $\mathcal{E}$ have a unique orthogonal projection, if we add the requirement of minimum distance, with the only exception of points on the equatorial disk

$$\left\{\rho_0 \leq e^2 a \, , \; z = 0\right\} .$$

References

Grafarend E.W. (1975). *Cartan frames and a foundation of physical geodesy*. In: Brosowski B., Martensen E. (eds), *Methoden und Verfahren der Mathematischen Physic*, 12:179–208, BI-Verlag, Mannheim.

Heiskanen W.A. and Moritz H. (1967). *Physical geodesy*. Freeman, San Francisco.

Sansò F., Sideris M.G. (2013). *Geoid determination: Theory and methods*. Lecture Notes in Earth System Sciences, Vol. 110. Springer-Verlag, Berlin, Heidelberg.

Chapter 3
The Earth Gravity Field: Basics

3.1 Outline

In this chapter we try to outline the main concepts used to estimate and describe the gravity field. The aim is to show the interplay between the geometry of the field, represented in terms of equipotential surfaces and plumb lines, and the mathematical relations that connect observable gravity values to the gravity potential. This is especially done in a linearized form, after a normal potential is defined, based on the ellipsoidal geometry, and used as reference function in the subsequent linearization.

In particular the knowledge of the gravity potential will prove to be essential to set up the transformation equation between different coordinate systems, as it will be done in Chap. 5.

3.2 Basic Definitions of Gravity and Gravity Potential

A massive, extended body B, with mass density $\rho(\mathrm{Q})$, considered in an inertial frame, exerts on a proof mass m, placed at point P, a force given by Newton's law

$$\boldsymbol{F}_{\mathrm{N}} = -Gm \int_B \frac{\boldsymbol{r}_{\mathrm{QP}}}{r_{\mathrm{QP}}^3}\, \rho(\mathrm{Q})\, \mathrm{d}B_{\mathrm{Q}}\,; \tag{3.1}$$

G is the universal gravitational constant. Since $\boldsymbol{F}_{\mathrm{N}}$ is just proportional to the proof mass m, if we divide the former by it we obtain a field of accelerations, the shape of which depends only on the mass distribution.

Therefore we define the gravitational (Newtonian) field $\boldsymbol{g}_{\mathrm{N}}$ as

$$\frac{\boldsymbol{F}_{\mathrm{N}}}{m} = \boldsymbol{g}_{\mathrm{N}} = -G \int_B \frac{\boldsymbol{r}_{\mathrm{QP}}}{r_{\mathrm{QP}}^3}\, \rho(\mathrm{Q})\, \mathrm{d}B_{\mathrm{Q}}\,. \tag{3.2}$$

F. Sansò et al., *Geodetic Heights*, Springer Geophysics,
https://doi.org/10.1007/978-3-030-10454-2_3

Note that $\boldsymbol{r}_{\mathrm{QP}}$ is the vector going from Q to P and that such formulas retain their meaning, wherever is P, outside or inside B, under some reasonable regularity conditions on $\rho\,(\mathrm{Q})$; for instance this holds if $\rho\,(\mathrm{Q})$ is measurable and bounded, as it is the true density of the masses in the Earth.

Let us immediately note that, due to the well known differential identity

$$\nabla_{\mathrm{P}} \frac{1}{r_{\mathrm{QP}}} = -\frac{\boldsymbol{r}_{\mathrm{QP}}}{r_{\mathrm{QP}}^3} \tag{3.3}$$

one gets from (3.2)

$$\boldsymbol{g}_{\mathrm{N}}\,(\mathrm{P}) = \nabla_{\mathrm{P}} \left(G \int\limits_B \frac{\rho\,(\mathrm{Q})}{r_{\mathrm{QP}}}\, \mathrm{d}B_{\mathrm{Q}} \right) , \tag{3.4}$$

meaning that $\boldsymbol{g}_{\mathrm{N}}$ is a potential field, i.e.

$$\boldsymbol{g}_{\mathrm{N}}\,(\mathrm{P}) = \nabla_{\mathrm{P}} V_{\mathrm{N}}\,(\mathrm{P}) \ , \qquad V_{\mathrm{N}}\,(\mathrm{P}) = G \int\limits_B \frac{\rho\,(\mathrm{Q})}{r_{\mathrm{QP}}}\, \mathrm{d}B_{\mathrm{Q}} \, . \tag{3.5}$$

The function $V_{\mathrm{N}}\,(\mathrm{P})$ is the gravitational potential generating $\boldsymbol{g}_{\mathrm{N}}$.

Now assume that the point P, where is the proof mass m, is sticked to a reference system (x, y, z) uniformly rotating around the z axis with angular velocity ω, as it happens on the Earth when P is kept fixed in a terrestrial reference frame. Without disturbing the more complicated Coriolis theory, we know that in addition to the gravitational attraction of B, m feels an apparent force, called centrifugal force $\boldsymbol{F}_{\mathrm{c}}$,

$$\boldsymbol{F}_{\mathrm{c}} = m\,\omega^2 \rho\, \boldsymbol{e}_\rho \, ; \tag{3.6}$$

recall that $\boldsymbol{e}_\rho$ is the unit radial vector in the equatorial plane and that

$$\rho\, \boldsymbol{e}_\rho = \rho \left(\cos\lambda\, \boldsymbol{e}_x + \sin\lambda\, \boldsymbol{e}_y \right) = x\, \boldsymbol{e}_x + y\, \boldsymbol{e}_y \, . \tag{3.7}$$

Therefore we can define a gravity field as the acceleration field, in the terrestrial reference system, given by

$$\begin{aligned} &\boldsymbol{g}\,(\mathrm{P}) = \boldsymbol{g}_{\mathrm{N}}\,(\mathrm{P}) + \frac{\boldsymbol{F}_{\mathrm{c}}}{m} = \boldsymbol{g}_{\mathrm{N}}\,(\mathrm{P}) + \boldsymbol{g}_{\mathrm{c}}\,(\mathrm{P}) \\ &\left(\boldsymbol{g}_{\mathrm{c}}\,(\mathrm{P}) = \omega^2 \rho\, \boldsymbol{e}_\rho = \omega^2 \left(x\, \boldsymbol{e}_x + y\, \boldsymbol{e}_y \right) \right) . \end{aligned} \tag{3.8}$$

Since the obvious relation holds

$$\boldsymbol{g}_{\mathrm{c}}\,(\mathrm{P}) = \nabla V_{\mathrm{c}}\,(\mathrm{P}) = \nabla \frac{1}{2} \omega^2 \left(x^2 + y^2 \right) = \nabla \left(\frac{1}{2} \omega^2 \rho^2 \right) . \tag{3.9}$$

we find that also $\boldsymbol{g}$ (P) is a potential field, namely that

$$\boldsymbol{g}\,(\mathrm{P}) = \nabla W = \nabla\left(V_{\mathrm{N}}\,(\mathrm{P}) + \frac{1}{2}\,\omega^2\rho^2\right)\ ; \tag{3.10}$$

the function W is called the gravity potential of the Earth.

We recall that a potential field is also irrotational and conservative, namely one has

$$\nabla \wedge \boldsymbol{g} \equiv 0 \tag{3.11}$$

as well as

$$\int_L \boldsymbol{g}\,(\mathrm{P}) \cdot \mathrm{d}\ell_{\mathrm{P}} \equiv 0$$

for any regular close line. The two facts depend one upon the other Hotine (1969), Marussi (1985).

One could object that given the physical field $\boldsymbol{g}$, its potential W is defined up to an arbitrary constant. However this ambiguity is cancelled when we choose the version

$$W\,(\mathrm{P}) = V_{\mathrm{N}}\,(\mathrm{P}) + \frac{1}{2}\,\omega^2\rho^2 \tag{3.12}$$

because V_{N} (P), as a consequence of its definition (3.5), has the unique property that

$$\lim_{r_{\mathrm{P}}\to\infty} V_{\mathrm{N}}\,(\mathrm{P}) = 0\ ; \tag{3.13}$$

one says that V_{N} (P) is regular at infinity.

Concerning this point, we study already here the asymptotic behaviour of V_{N} (P) when $r_{\mathrm{P}} \to \infty$. Assume r_{P} to be larger than the Brillouin radius

$$R_+ = \max_{\mathrm{Q}\in B} r_{\mathrm{Q}}$$

so that

$$\frac{r_{\mathrm{Q}}}{r_{\mathrm{P}}} < \frac{R_+}{r_{\mathrm{P}}} < 1$$

for every Q in B, then the following elementary relations hold

$$\begin{aligned}\frac{1}{\left|\boldsymbol{r}_{\mathrm{P}} - \boldsymbol{r}_{\mathrm{Q}}\right|} &= \frac{1}{\sqrt{r_{\mathrm{P}}^2 + r_{\mathrm{Q}}^2 - 2r_{\mathrm{P}}r_{\mathrm{Q}}\,\boldsymbol{e}_{\mathrm{P}}\cdot\boldsymbol{e}_{\mathrm{Q}}}} = \\ &= \frac{1}{r_{\mathrm{P}}}\,\frac{1}{\sqrt{1 + \left(\dfrac{r_{\mathrm{Q}}}{r_{\mathrm{P}}}\right)^2 - 2\left(\dfrac{r_{\mathrm{Q}}}{r_{\mathrm{P}}}\right)\boldsymbol{e}_{\mathrm{P}}\cdot\boldsymbol{e}_{\mathrm{Q}}}}\end{aligned}$$

$$= \frac{1}{r_P}\left[1 + \frac{r_Q}{r_P}\boldsymbol{e}_P \cdot \boldsymbol{e}_Q + \mathcal{O}_2\right] \tag{3.14}$$

where $\boldsymbol{e}_P = \frac{\boldsymbol{r}_P}{r_P}$ and $\boldsymbol{e}_Q = \frac{\boldsymbol{r}_Q}{r_Q}$.

In (3.14) $\mathcal{O}_2$ means a second order term in $\frac{r_Q}{r_P}$. So we can write

$$V_N(P) = \frac{G}{r_P}\int_B \rho(Q)\, dB_Q + G\frac{\boldsymbol{e}_P}{r_P^2} \cdot \int_B \rho(Q)\, r_Q \boldsymbol{e}_Q\, dB_Q + \mathcal{O}_3\,, \tag{3.15}$$

where $\mathcal{O}_3$ is a third order infinitesimal in $\frac{1}{r_P}$.

Noting that

$$\int_B \rho(Q)\, dB_Q = M\,, \tag{3.16}$$

the mass of the Earth, and that

$$\frac{1}{M}\int_B \rho(Q)\, r_Q \boldsymbol{e}_Q\, dB_Q = \frac{1}{M}\int_B \rho(Q)\, \boldsymbol{r}_Q\, dB_Q = \boldsymbol{b}\,, \tag{3.17}$$

the barycentre of the Earth masses, (3.15) writes, putting $\mu = GM$,

$$V_N(P) = \frac{\mu}{r_P} + G\frac{\boldsymbol{r}_P}{r_P^3} \cdot \boldsymbol{b} + \mathcal{O}_3\,. \tag{3.18}$$

We have already seen that choosing z as the rotation axis considerably simplifies the expression of the centrifugal force (3.6), so now a further clever choice of the origin of (x, y, z), that we keep fixed from now on, will simplify (3.18); namely we put the origin O of our reference system at the barycentre of the masses, so that $\boldsymbol{b} = 0$ and (3.18) becomes

$$V_N(P) = \frac{\mu}{r_P} + \mathcal{O}_3\,. \tag{3.19}$$

We formulate now a fundamental differential property of $V_N(P)$. One can prove in fact (see Sansò and Sideris 2013) that

$$\Delta V_N(P) = -4\pi G\rho(P) \equiv \begin{cases} -4\pi G\rho(P) \neq 0 & \text{(inside } B) \\ 0 & \text{(outside } B) \end{cases}\,. \tag{3.20}$$

In particular $V_N(P)$ is a harmonic function in Ω, the space outside the masses. We note as well that from (3.20) and (3.12) descends the relation

$$\Delta W = -4\pi G\rho(P) + 2\omega^2\,. \tag{3.21}$$

It is important however to underline that while V_N is the unique solution of (3.20) being also regular at infinity, W cannot be considered as "the" solution of (3.21); (3.21) should always be specified as the sum of V_N and the centrifugal potential $\frac{1}{2}\omega^2\left(x^2+y^2\right)$.

3.3 Plumb Lines and Equipotential Surfaces

First of all we define the field of the physical vertical directions.

This is a vector field, $\boldsymbol{n}$ (P), with modulus 1, with the same direction and opposite verse of $\boldsymbol{g}$ (P), namely

$$\boldsymbol{n}(\mathrm{P}) = -\frac{\boldsymbol{g}(\mathrm{P})}{g(\mathrm{P})} . \tag{3.22}$$

The importance of this vector field stems from the fact that there are many geodetic instruments capable of materializing the vector $\boldsymbol{n}$ at any point in free-space and therefore many geodetic observations refer to it (e.g. zenith distances, levelling increments, etc.). As a matter of fact, $\boldsymbol{n}$ can also be determined with respect to a celestial reference system (the system of so called fixed stars) by means of astro-geodetic observations and then rotated into the terrestrial system. This procedure however is not very commonly applied due to the length of the measurement operations (Vanìcek and Krakiwsky 1986).

Of great importance for us however is the family of force-lines of $\{\boldsymbol{n}(\mathrm{P})\}$, namely the so called plumb lines or vertical lines, that admit $\boldsymbol{n}$ (P) as tangent field. Such lines can have a cumbersome behaviour when we go a considerable distance out of the Earth, because there the centrifugal term can become large with respect to the Newtonian attraction. In fact on the equatorial plane they balance at about 37,700 km from the planet. Yet, in the layer of interest for us, the centrifugal acceleration is of the order of $3 \cdot 10^{-3}$ times the Newtonian gravity, which then dictates the shape of plumb lines.

Calling $\{L_v\}$ the family of vertical lines, we note that in our relevant layer they never cross one another, i.e. through every point P passes one and only one line L_v.

This implies also that along each L_v one can introduce an arc-length parameter that is always increasing upward, namely in the direction of $\boldsymbol{n}$ (P). The line differential equation in the intrinsic arc-length parameter s is then

$$\frac{\mathrm{d}\boldsymbol{r}(\mathrm{P})}{\mathrm{d}s} = \boldsymbol{n}(\mathrm{P}) . \tag{3.23}$$

Having defined the family of lines $\{L_v\}$, one can consider the family of surfaces

$$S_{\overline{W}} = \left\{\mathrm{P}\,;\;\; W(\mathrm{P}) = \overline{W}\right\} , \tag{3.24}$$

which are called equipotential surfaces, each corresponding to some constant gravity potential value $\overline{W}$.

The relevant values of $\overline{W}$, corresponding to equipotentials close to the Earth surface, are around $\overline{W} \cong 6.3710 \cdot 10^7\ \mathrm{m}^2\mathrm{s}^{-2}$ plus minus a relative variation of $0.5 \cdot 10^{-2}$. One can prove that the equipotential surfaces are so regular as to admit a normal field, and that they can never cross one another. Moreover they are closed surfaces in the topographic layer and so they are naturally ordered from inside to outside the masses.

The most relevant property of the two families $\{L_\upsilon\}$ and $\left\{S_{\overline{W}}\right\}$ is contained in the following elementary proposition.

Proposition 3.1 *Given an equipotential $S_{\overline{W}}$, its (exterior) normal field corresponds to $\left\{\boldsymbol{n}(P)\,;\ P \in S_{\overline{W}}\right\}$; so the family $\{L_\upsilon\}$ crosses $S_{\overline{W}}$ orthogonally.*

This is immediate, in fact if $\mathrm{P} \in S_{\overline{W}}$ and we move away from P tangentially to $S_{\overline{W}}$, we have indeed $\mathrm{d}W = 0$, because

$$W(\mathrm{P}) = \overline{W}\,, \quad \mathrm{P} \in S_{\overline{W}}\,;$$

but then the differential relation

$$\mathrm{d}W = \boldsymbol{g}(\mathrm{P}) \cdot \mathrm{d}\boldsymbol{r}$$

implies

$$\mathrm{d}\boldsymbol{r} \text{ tangent to } S_{\overline{W}} \Rightarrow \mathrm{d}\boldsymbol{r} \perp \boldsymbol{g}\,;\ \mathrm{d}\boldsymbol{r} \perp \boldsymbol{n}\,,$$

i.e. $\boldsymbol{n}$ is normal to $S_{\overline{W}}$ in P.

Other geometric quantities that describe in a more subtle way the behaviour of $\{L_\upsilon\}$ and $\left\{S_{\overline{W}}\right\}$ are respectively the principal curvature vector $\boldsymbol{c}$ of the line L_υ and the mean curvature $\mathcal{C}$. These are defined as (Hotine 1969)

$$\boldsymbol{c}(\mathrm{P}) = \frac{\mathrm{d}\boldsymbol{n}}{\mathrm{d}\ell}\,, \tag{3.25}$$

where ℓ is the arc-length of the plumb line and

$$\mathcal{C}(\mathrm{P}) = \frac{1}{2}\left(\frac{1}{R_1} + \frac{1}{R_2}\right)\,, \tag{3.26}$$

where R_1, R_2 are the minimum and maximum radius of curvature of the normal sections of $S_{\overline{W}}$ at P.

Such quantities are related to the variability of the modulus $g(\mathrm{P})$ in space. In particular, if we call ∇_0 the gradient along the surface $S_{\overline{W}}$, one has

$$\boldsymbol{c} = \frac{\nabla_0 g}{g}\,; \tag{3.27}$$

on the contrary, taking the derivative of g along L_v, one gets the relation

$$\frac{dg}{d\ell} = -2\mathcal{C}g + 4\pi G\rho - 2\omega^2 . \tag{3.28}$$

The Eq. (3.28) holds both inside and outside the masses, where ρ is equal to zero. The proof of above equations can be found e.g. in (Sansò and Sideris 2013, Part I, Sect. 1.7).

Before closing the section we need a quantitative assessment of the degree of parallelism of equipotential surfaces. If we move vertically from a point A on the equipotential surface S_{W_A} to a point A′ on the equipotential surface $S_{W_A+\delta W}$, we see that the rough relation $\delta W = -g_A L_A$ holds, where L_A denotes the vertical distance between A and A′. If we repeat the same operation between two other points B, B′ on the same surfaces, we will find $\delta W = -g_B L_B$. These relations imply

$$\frac{L_B - L_A}{L_A} = -\frac{g_B - g_A}{g_A} .$$

Since the maximum horizontal variation of g (from pole to equator) is of the order of $|\delta g| \sim 5 \cdot 10^{-3}\, g$, we see that we have too $|\delta L| \sim 5 \cdot 10^{-3}\, L$. With L up to 2 m, this lack of parallelism accounts to a global maximum of 1 cm, and can therefore be neglected. Note however that with $L = 100$ m, $\delta L = 0.5$ m and this is not negligible anymore. Of course at the level of high mountains the parallelism is lost.

3.4 The Gravity Field Outside a Brillouin Sphere

Let us recall that we have defined a minimal Brillouin sphere as the one centered at the origin with a radius R_+,

$$R_+ = \max_{P \in B} r_P ; \tag{3.29}$$

a Brillouin sphere is a sphere with radius $R > R_+$, so that all the masses generating the gravity field are at its interior, i.e.

$$\frac{r_P}{R} \le \frac{R_+}{R} < 1 , \quad P \in B . \tag{3.30}$$

Given this condition we want to develop a representation of the gravitational part of the gravity field, namely V_N (P), on and outside a Brillouin sphere. This will be achieved by pushing the reasoning of Sect. 3.2 to a full series representation of the Newton kernel r_{QP}^{-1}.

To this purpose we prepare a proposition of purely algebraic nature.

Proposition 3.2 *Consider the function of two variables*

$$G(s,t) = \frac{1}{\sqrt{1+s^2-2st}}, \tag{3.31}$$

also called Legendre generating function, in the set

$$0 \le s < 1; \quad -1 \le t \le 1 . \tag{3.32}$$

$G(s,t)$ *is real analytic in* s $(s < 1)$, *for every fixed* t, *so that the convergent series*

$$G(s,t) = \sum_{n=0}^{+\infty} s^n P_n(t) \tag{3.33}$$

holds. It turns out that (Abramowitz and Stegun 1964; Heiskanen and Moritz 1967; Sansò and Sideris 2013):

- $P_n(t)$ are polynomials of degree n and with the same parity (even or odd) as n, called Legendre polynomials,
- $P_0(t) \equiv 1$, $P_1(t) \equiv t$, and that the higher degree $P_n(t)$ satisfy the recursive relation

$$(n+1)\, P_{n+1}(t) \equiv (2n+1)\, t P_n(t) - n P_{n-1}(t) \ , \tag{3.34}$$

- $P_n(1) = 1$, $P_n(-1) = (-1)^n$ and

$$|P_n(t)| < 1 \quad -1 < t < 1 , \tag{3.35}$$

- $\{P_n(t)\}$ is an orthogonal system in $L^2(-1,1)$, namely

$$\int_{-1}^{1} P_n(t)\, P_j(t)\, \mathrm{d}t = \frac{2\,\delta_{nj}}{2n+1}, \tag{3.36}$$

- $\{P_n(t)\}$ is complete in $L^2(-1,1)$, i.e. $\forall f(t) \in L^2(-1,1)$ the following identity holds

$$f(t) = \sum_{n=0}^{+\infty} \frac{2n+1}{n} P_n(t) \int_{-1}^{1} P_n(\tau)\, f(\tau)\, \mathrm{d}\tau ; \tag{3.37}$$

  the convergence of (3.37) is in the $L^2(-1,1)$ topology, i.e. setting

$$f_N(t) = \sum_{n=0}^{N} \frac{2n+1}{n} P_n(t) \int_{-1}^{1} P_n(\tau)\, f(\tau)\, \mathrm{d}\tau ,$$

one has

$$\lim_{N\to+\infty}\int_{-1}^{1}\left[f\left(t\right)-f_{N}\left(t\right)\right]^{2}\mathrm{d}t=0\,. \tag{3.38}$$

We exploit now Proposition 3.2 to compute a series representation of $V_{\mathrm{N}}\left(\mathrm{P}\right)$. Note that, calling

$$\cos\psi_{\mathrm{PQ}}=\frac{\boldsymbol{r}_{\mathrm{P}}\cdot\boldsymbol{r}_{\mathrm{Q}}}{r_{\mathrm{P}}\,r_{\mathrm{Q}}}\,,$$

and assuming that P is on or outside a Brillouin sphere, while Q is in B, the following relation holds

$$\begin{aligned}\frac{1}{r_{\mathrm{QP}}}&=\frac{1}{r_{\mathrm{P}}}\frac{1}{\sqrt{1+\left(\frac{r_{\mathrm{Q}}}{r_{\mathrm{P}}}\right)^{2}-2\left(\frac{r_{\mathrm{Q}}}{r_{\mathrm{P}}}\right)\cos\psi_{\mathrm{PQ}}}}=\\&=\frac{1}{r_{\mathrm{P}}}\,G\left(\frac{r_{\mathrm{Q}}}{r_{\mathrm{P}}},\cos\psi_{\mathrm{PQ}}\right)=\\&=\sum_{n=0}^{+\infty}\frac{r_{\mathrm{Q}}^{n}}{r_{\mathrm{P}}^{n+1}}\,P_{n}\left(\cos\psi_{\mathrm{PQ}}\right)\;;\end{aligned} \tag{3.39}$$

the series is convergent even uniformly in r_{Q} because

$$r_{\mathrm{Q}}\leq R_{+}<R\leq r_{\mathrm{P}}\,.$$

By using (3.39) in (3.5) we get

$$V_{\mathrm{N}}\left(\mathrm{P}\right)=G\sum_{n=0}^{+\infty}\frac{1}{r_{\mathrm{P}}^{n+1}}\int_{B}r_{\mathrm{Q}}^{n}\,P_{n}\left(\cos\psi_{\mathrm{PQ}}\right)\rho\left(\mathrm{Q}\right)\mathrm{d}B_{\mathrm{Q}}\,. \tag{3.40}$$

To avoid having terms of different physical dimension for each n, we rewrite (3.40) in the form

$$V_{\mathrm{N}}\left(\mathrm{P}\right)=\frac{GM}{R}\sum_{n=0}^{+\infty}\left(\frac{R}{r_{\mathrm{P}}}\right)^{n+1}\left(\frac{1}{M}\int_{B}\left(\frac{r_{\mathrm{Q}}}{R}\right)^{n}P_{n}\left(\cos\psi_{\mathrm{PQ}}\right)\rho\left(\mathrm{Q}\right)\mathrm{d}B_{\mathrm{Q}}\right)\,, \tag{3.41}$$

with M the mass of the Earth. Note that $\frac{GM}{R}$ has the dimension of a gravitational potential, while all the other terms are a-dimensional.

Now we have to introduce a Lemma that is fundamental for harmonic calculus in spherically symmetric domains.

Lemma 3.1 (the summation rule) *The following identity holds*

$$P_n\left(\cos\psi_{PQ}\right)=\frac{1}{2n+1}\sum_{m=-n}^{n}Y_{nm}\left(\sigma_P\right)Y_{nm}\left(\sigma_Q\right) \tag{3.42}$$

$$\left(\sigma_P\equiv\left(\lambda_P,\varphi_P\right),\ \sigma_P\equiv\left(\lambda_Q,\varphi_Q\right)\right)\ ,$$

where the functions $Y_{nm}\left(\sigma_P\right)$, called spherical harmonics of degree n and order m, are given by

$$Y_{nm}(\sigma)=Y_{nm}(\lambda,\varphi)=\overline{P}_{nm}(\varphi)\begin{cases}\cos m\lambda & (m\geq 0)\\ \sin m\lambda & (m<0)\end{cases} \tag{3.43}$$

and $\overline{P}_{nm}(\varphi)$ are the so called associated normalized Legendre functions, given by

$$\overline{P}_{nm}(\varphi)=(\cos\varphi)^m\left[D_t^m P_n(t)\right]_{t=\sin\varphi}\sqrt{(2-\delta_{n0})\frac{(2n+1)(n-m)!}{(n+m)!}}\ . \tag{3.44}$$

We list here some properties of the associated Legendre functions and of the spherical harmonics (proofs can be found in Sansò and Sideris 2013, Sect. 3.4):

- putting $t=\sin\varphi$,

$$\overline{P}_{m-1,m}(t)\equiv 0\ ,\quad \overline{P}_{m,m}(t)=\sqrt{\frac{2(2m+1)}{(2m)!}}(\cos\varphi)^m \tag{3.45}$$

- $\overline{P}_{nm}(t)$ satisfy the following recursive relation

$$\begin{aligned}\overline{P}_{n+1,m}(t)=&\sqrt{\frac{(2n+1)(2n+3)}{(n+1-m)(n+1+m)}}\,t\,\overline{P}_{nm}(t)+\\ &-\sqrt{\frac{(2n+3)\left(n^2-m^2\right)}{(2n-1)(n+1-m)(n+1+m)}}\overline{P}_{n-1,m}(t)\ ,\end{aligned} \tag{3.46}$$

- $\{Y_{nm}(\sigma)\}$ is an orthonormal system in $L^2(\sigma)$, i.e.

$$\frac{1}{4\pi}\int Y_{nm}(\sigma)Y_{jk}(\sigma)\,\mathrm{d}\sigma=\delta_{nj}\delta_{mk} \tag{3.47}$$

$$(\mathrm{d}\sigma=\cos\varphi\,\mathrm{d}\varphi\,\mathrm{d}\lambda)\ ,$$

- $\{Y_{nm}(\sigma)\}$ is a complete system in $L^2(\sigma)$, i.e. $\forall f\in L^2(\sigma)$ the following identity holds

$$f(t) = \sum_{n=0}^{+\infty} \sum_{m=-n}^{n} Y_{nm}(\sigma) \left(\frac{1}{4\pi} \int Y_{nm}(\sigma') f(\sigma') \, d\sigma' \right) , \tag{3.48}$$

the convergence of the series being in the sense of $L^2(\sigma)$.

Equipped with the above discussion, we can finally obtain the main result of this section, namely the decomposition of the Newton integral for $V_N(P)$ into a series of functions depending on P only multiplied by integrals on the variable $Q \in B$. In fact, substituting (3.42) into (3.41) and setting $\mu = GM$, we find

$$V_N(P) = \frac{\mu}{R} \sum_{n=0}^{+\infty} \sum_{m=-n}^{n} \left(\frac{R}{r_P} \right)^{n+1} \frac{Y_{nm}(\sigma_P)}{(2n+1)} \left(\frac{1}{M} \int_B \rho(Q) \left(\frac{r_Q}{R} \right)^n Y_{nm}(\sigma_Q) \, dB_Q \right) \equiv$$
$$\equiv \sum_{n=0}^{+\infty} \sum_{m=-n}^{n} S_{nm}(r_P, \sigma_P) \, V_{nm} , \tag{3.49}$$

where

$$S_{nm}(r_P, \sigma_P) = \left(\frac{R}{r_P} \right)^{n+1} Y_{nm}(\sigma_P) \tag{3.50}$$

are called exterior solid spherical harmonics, and

$$V_{nm} = \frac{1}{2n+1} \frac{\mu}{R} \frac{1}{M} \int_B \rho(Q) \left(\frac{r_Q}{R} \right)^n Y_{nm}(\sigma_Q) \, dB_Q . \tag{3.51}$$

Let us observe that (3.49) tells us that V_N is a linear combination of $S_{nm}(r, \sigma)$ and that such functions are linearly independent from one another, as we can see from (3.47); so, since V_N has to be harmonic outside S_R, the same must be true for $S_{nm}(r, \sigma)$, whence their name.

Remark (Poisson function theory) First of all we note that on the Brillouin sphere, $r_P = R$, we have

$$S_{nm}(R, \sigma_P) = Y_{nm}(\sigma_P) ; \tag{3.52}$$

since it is easy to prove that $S_{nm}(r, \sigma)$ are harmonic functions in $r > R$, we see that $S_{nm}(r, \sigma)$ are characterized as the harmonic functions outside the Brillouin sphere that on it coincide with the spherical harmonics $Y_{nm}(\sigma)$.

A second remark now is that $V_N(P)$ by (3.49) can be represented as

$$V_N(P) = \sum_{n=0}^{+\infty} \sum_{m=-n}^{n} V_{nm} \, S_{nm}(r_P, \sigma_P)$$

and, on the sphere of radius R, one has

$$V_{\mathrm{N}}(R, \sigma_{\mathrm{P}}) = \sum_{n=0}^{+\infty} \sum_{m=-n}^{n} V_{nm} S_{nm}(\sigma_{\mathrm{P}}) \; . \tag{3.53}$$

Now note that we can take for $V_{\mathrm{N}}(R, \sigma_{\mathrm{P}})$ any $L^2(\sigma)$ function and, recalling (3.48), we know that the series (3.53) holds with

$$V_{nm} = \frac{1}{4\pi} \int V_{\mathrm{N}}\left(R, \sigma'\right) Y_{nm}\left(\sigma'\right) \mathrm{d}\sigma' \; . \tag{3.54}$$

This means that assigning the values of $V_{\mathrm{N}}(\mathrm{P})$ on the sphere of radius R, we can find the function, harmonic outside the sphere, that agrees with them on the sphere (Dirichlet problem).

Said in another way, the harmonic function $V_{\mathrm{N}}(\mathrm{P})$ in space is fixed once its values on a Brillouin sphere are given, irrespectively of what is the internal mass distribution $\rho(\mathrm{Q})$. In particular one can follow the reverse way to find

$$\begin{aligned} V_{\mathrm{N}}(r, \sigma) &= \sum_{n=0}^{+\infty} \sum_{m=-n}^{n} \left(\frac{R}{r_{\mathrm{P}}}\right)^{n+1} Y_{nm}(\sigma_{\mathrm{P}}) \left(\frac{1}{4\pi} \int V_{\mathrm{N}}\left(R, \sigma'\right) Y_{nm}\left(\sigma'\right) \mathrm{d}\sigma'\right) = \\ &= \frac{1}{4\pi} \int \sum_{n=0}^{+\infty} \left(\frac{R}{r_{\mathrm{P}}}\right)^{n+1} (2n+1) P_n\left(\cos\psi_{\mathrm{PQ}}\right) V_{\mathrm{N}}(\mathrm{Q}) \, \mathrm{d}\sigma_{\mathrm{Q}} = \\ &= \int \Pi\left(r_{\mathrm{P}}, \sigma_{\mathrm{P}}, \sigma_{\mathrm{Q}}\right) V_{\mathrm{N}}(\mathrm{Q}) \, \mathrm{d}\sigma_{\mathrm{Q}} \; , \end{aligned} \tag{3.55}$$

where the sum of the series is the Poisson function, which is explicitly given by

$$\Pi\left(r_{\mathrm{P}}, \sigma_{\mathrm{P}}, \sigma_{\mathrm{Q}}\right) = \frac{1}{4\pi} \frac{R\left(r_{\mathrm{P}}^2 - R^2\right)}{r_{\mathrm{QP}}^3} \; . \tag{3.56}$$

3.5 The Normal Gravity Field

This is a purely mathematical field, constructed to provide a suitable approximation to the actual gravity field of the Earth.

One could think that a good approximation is already obtained by adding the centrifugal potential to a purely spherical potential, i.e.

$$W_{\mathrm{S}}(\mathrm{P}) = \frac{\mu}{r_{\mathrm{P}}} + \frac{1}{2}\omega^2 r_{\mathrm{P}}^2 \cos^2\varphi_{\mathrm{P}} \; .$$

This approximation however is known on an experimental basis to be too rough, implying a relative error in the potential of the order of 10^{-3}. The point is that the Earth is not a rigid body, nor a purely elastic one, and its rotation has created a permanent flattening of its shape, contracting the polar radius and dilating the equatorial radius.

So the approximate shape of the Earth, and then its gravity field, should account for this ellipsoidal geometry. This geometry in fact is well suited to approximate equipotential surfaces of W; it is known that, by a careful choice of the parameters of an ellipsoid, we can approximate an equipotential with a relative error of the order of 10^{-5}, which is certainly useful for a subsequent linearization.

The precise definition of normal gravity potential is as follows: let us take an oblate ellipsoid $\mathcal{E}$, the shape of which is fixed by two parameters, the equatorial radius a and the eccentricity e; place this $\mathcal{E}$ with the centre at the barycentre of the masses and the polar (short) axis along the rotation axis z; we define then a normal gravity potential U (P) that is composed by a harmonic part, V_e (P), regular at infinity, and a centrifugal part $V_c\,(P) = \frac{1}{2}\omega^2\rho^2$, as the centrifugal potential contained in W (P). $\mathcal{E}$ has to be an equipotential of U (P).

Such a definition is sufficient to uniquely identify V_e (P) and then U (P) too, apart from the constant value $U_0 = U\,(P)|_{\mathcal{E}}$. In fact V_e (P) has to satisfy the Dirichlet problem

$$\begin{cases} \Delta V_e = 0 & \text{outside } \mathcal{E} \\ V_e|_{\mathcal{E}} = U_0 - \frac{1}{2}\omega^2\left(x^2+y^2\right) = U_0 - \frac{1}{2}\omega^2\mathcal{N}^2\cos^2\varphi \\ V_e \to 0 & \text{at infinity} \end{cases} . \tag{3.57}$$

Not only (3.57) has one and only one solution, but even, expressing the problem in a suitable system of ellipsoidal coordinates, one can find its exact analytic expression (see Pizzetti 1894, Somigliana 1929, 1930 and Sansò and Sideris 2013 for an elementary derivation).

Even without writing it, we know in advance that V_e will depend only on h and φ and not on λ; this is due to the obvious cylindrical symmetry of (3.57). Therefore also U, given by

$$U\,(h,\varphi) = V_e\,(h,\varphi) + \frac{1}{2}\omega^2\,(\mathcal{N}+h)^2\cos^2\varphi\,, \tag{3.58}$$

results to be independent from λ.

Now we can define the normal gravity vector as

$$\boldsymbol{\gamma}\,(h,\varphi) = \nabla U\,(h,\varphi)\ ; \tag{3.59}$$

we note that since U is not a function of λ, the normal vector $\boldsymbol{\gamma}$ has no component on $\boldsymbol{e}_\lambda$. Moreover, since $\mathcal{E}$ is an equipotential of U, then $\boldsymbol{\gamma}$ is orthogonal to $\mathcal{E}$, more precisely it has the same direction and opposite verse of $\boldsymbol{\nu}$, the normal to $\mathcal{E}$.

In the sequel we will also need the normal gravity modulus

$$\gamma\,(h,\varphi) = |\boldsymbol{\gamma}\,(h,\varphi)| = |\nabla U\,(h,\varphi)|\ . \tag{3.60}$$

Table 3.1 Parameters of the GRS80 ellipsoidal field

$\mu = GM = 398,600.5\ \text{km}^3\text{s}^{-2}$
$a = 6,378.137$ km
$e^2 = 6.69438002290 \cdot 10^{-3}$
$\gamma_0 = 978.03267715$ Gal

Note that we will use Gal $=$ cm s^{-2} as measurement unit for gravity.

We shall also use the notation $\widetilde{\boldsymbol{n}}$ to indicate the normal vertical unit vector, namely

$$\widetilde{\boldsymbol{n}}\,(h,\varphi) = -\frac{\boldsymbol{\gamma}\,(h,\varphi)}{\gamma\,(h,\varphi)}\,; \tag{3.61}$$

as it is obvious $\widetilde{\boldsymbol{n}}$ coincides with $\boldsymbol{\nu}$ when $h = 0$, i.e. on $\mathcal{E}$.

Instead of the closed expression of U and γ, we go here with the traditional formulas of the International Association of Geodesy (IAG), which express U and γ as functions of (λ, φ, h), because though approximate, they are quite easily applied in calculations. Such formulas are developments of potential and gravity in series of h, truncated to the second order, which guarantees an accuracy at the μGal level in the topographic layer.

The constants reported in Table 3.1 have been used in the calculations. From these we obtain the formulas reported in Table 3.2 and expressed in units of Gal's and km's. All numerical constants are referred to the GRS80 reference system (Moritz 1988).

Using the formulas in Table 3.2, one can in particular derive the following relation, valid up to 6 km, with a relative approximation better than 10^{-8},

$$\widetilde{\boldsymbol{n}} \cong \boldsymbol{\nu} + \frac{\gamma_{e\varphi}}{\gamma_0}\frac{h}{a}\,\boldsymbol{e}_\varphi = \boldsymbol{\nu} + 5.30244 \cdot 10^{-3} \sin 2\varphi \,\frac{h}{a}\,\boldsymbol{e}_\varphi\,. \tag{3.62}$$

Such an equation gives us two important pieces of information that will be used in the sequel.

With reference to Fig. 3.1, we see that, defining a normal deflection of the vertical $\widetilde{\delta}$ as the angle between $\widetilde{\boldsymbol{n}}$ and $\boldsymbol{\nu}$, one has

$$\widetilde{\delta} \cong |\widetilde{\boldsymbol{n}} - \boldsymbol{\nu}| \cong \frac{\gamma_{e\varphi}}{\gamma_0}\frac{h}{a}\,. \tag{3.63}$$

So computing $\widetilde{\delta}$ in the most unfavourable conditions, in the topographic layer, i.e. with $\varphi = 45°$ and $h = 6$ km, one has

$$\widetilde{\delta}_{\max} \leq 5 \cdot 10^{-6}\,, \tag{3.64}$$

corresponding to an arc-second.

Also interesting is to estimate the distance $\overline{PP'}$ of the normal vertical, $\widehat{\mathrm{P_eP'}}$ in Fig. 3.1, from the normal to $\mathcal{E}$; always in the worst case $\varphi = 45°$ and $h = 6$ km, one has

Table 3.2 Normal potential and normal gravity formulas (γ in Gals and h in km's)

γ = gravity modulus

$$\gamma(h,\varphi) = \gamma_e(\varphi) - \gamma_1(\varphi)\,h + \gamma_2(\varphi)\,h^2$$

$$\gamma_e(\varphi) = \gamma_0\left(1 + 5.30244\cdot 10^{-3}\sin^2\varphi - 5.8\cdot 10^{-6}\sin^2 2\varphi\right)$$

$$\gamma_1(\varphi) = 0.30877 - 4.5\cdot 10^{-4}\sin^2\varphi$$

$$\gamma_2(\varphi) = 72\cdot 10^{-6}$$

$\boldsymbol{\gamma}$ = gravity vector (v = vertical component, t = tangent component)

$$\boldsymbol{\gamma}(h,\varphi) = v(h,\varphi)\,\boldsymbol{\nu}(\varphi) + t(h,\varphi)\,\boldsymbol{e}_\varphi(\varphi)$$

$$v = -\gamma(h,\varphi) + \frac{1}{2\gamma_0}\,\tau_1^2\,h^2$$

$$t = \tau_1\,h + \tau_2\,h^2$$

$$\tau_1 = -\frac{\gamma_{e\varphi}}{\mathcal{M}} = -\frac{\gamma_0}{\mathcal{M}}\left(5.30244\cdot 10^{-3}\sin 2\varphi - 11.6\cdot 10^{-6}\sin 4\varphi\right)$$

$$\tau_2 = -\frac{1}{2\mathcal{M}}\left(\gamma_{1\varphi} + 2\frac{\gamma_{e\varphi}}{\mathcal{M}}\right) = -\frac{1}{2\mathcal{M}}\left(4.5\cdot 10^{-4}\sin 2\varphi + 2\tau_1\right)$$

$$\left(\gamma_{e\varphi} = \frac{\partial}{\partial\varphi}\gamma_e(\varphi)\,,\quad \gamma_{1\varphi} = \frac{\partial}{\partial\varphi}\gamma_1(\varphi)\right)$$

U = gravity potential

$$U(h,\varphi) = U_0 - \gamma_e(\varphi)\,h + \gamma_1(\varphi)\,\frac{1}{2}h^2 - \left(\gamma_2(\varphi) - \frac{1}{2}\frac{\tau_1^2}{\gamma_0}\right)\frac{1}{3}h^3$$

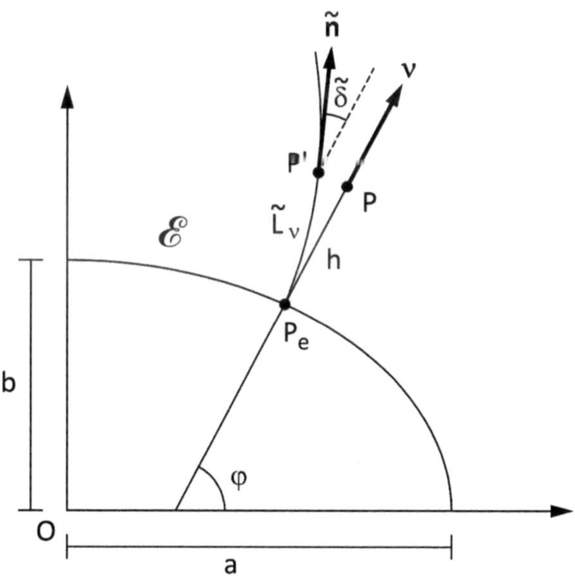

Fig. 3.1 The geometry of normal vertical lines with respect to the normal to $\mathcal{E}$; $\boldsymbol{\nu}$ normal to $\mathcal{E}$, $\widetilde{\boldsymbol{n}}$ normal vertical, $\widetilde{\delta}$ normal deflection of the vertical, $h = \overline{P_eP}$, $\widetilde{L}_v = \widehat{P_eP'}$

$$\overline{\mathrm{PP}'} = \left| \int_0^h \widetilde{\delta}\left(h'\right) \mathrm{d}h' \right| = \frac{1}{2} \frac{\gamma_{e\varphi}}{\gamma_0} \frac{h^2}{a} < 15 \cdot 10^{-6}\ \mathrm{km} = 15\ \mathrm{mm}. \tag{3.65}$$

Moreover the length $\widetilde{L}_v = \widehat{\mathrm{P_eP'}}$ can be compared to $h = \overline{\mathrm{P_eP}}$ by the relation

$$\left|\widetilde{L}_v - h\right| = \int_0^{\widetilde{L}_v} \left(\cos\widetilde{\delta} - 1\right) \mathrm{d}L_v \cong \int_0^h \frac{1}{2} \widetilde{\delta}^2\, \mathrm{d}h$$

so that, using (3.63) and integrating in dh,

$$\left|\widetilde{L}_v - h\right| \le \frac{1}{6} \widetilde{\delta}^2_{\max} h = 25 \cdot 10^{-12}\ \mathrm{km}\ , \tag{3.66}$$

which is practically zero in the frame of this text.

All in all we could conclude this discussion on the geometry of normal vertical lines by claiming that, in practice, *normal vertical lines and lines orthogonal to $\mathcal{E}$ can be considered as coincident in the topographic layer.*

Remark On the value of U_0.

Although we have not given the explicit formula for U_e, and hence for U, in ellipsoidal coordinates, we want to state here that such a formula implies the following relation, holding on $\mathcal{E}$

$$U_0 = \frac{\mu}{ea} \operatorname{arctg} \frac{e}{\sqrt{1 - e^2}} + \frac{1}{3}\omega^2 a^2 \tag{3.67}$$
$$(\mu = GM)\ .$$

Furthermore, by examining the asymptotic behaviour of V_e when $r \to \infty$, one can see that

$$V_e \sim \frac{\mu}{r} - \mu J_2 \frac{a^2}{r^3} P_2(\cos\vartheta) + \ldots\ , \tag{3.68}$$

where the term $\mathcal{O}\left(\frac{1}{r^2}\right)$ is missing because the origin is placed at the barycentre implied by the normal field. The point here is that, by tracking orbits of satellites flying in the potential (3.68), one can estimate μ and $\mu J_2 a^2$. The value of μ is the one already presented in Table 3.1. The value of a represents a scale factor for the whole gravimetric problem and, in any event, it has been conventionally fixed too, to the value reported in Table 3.1. Notice that the two mentioned values refer to the reference system GRS80, while today updated values are available, for instance those supported by the International Earth Rotation and Reference Systems Service (IERS), see the website www.iers.org. In any case we underline that, in spite of different arguments that have been raised in literature (see Martinec 1998), for the pure purpose of approximation theory that we pursue here, any value of a that does not produce an exit from the linearization band is suitable for our computations. In any case, once μ and a are fixed, from the satellite tracking again one can derive J_2 which results, adopting the values of Table 3.1,

$$J_2 = 0.00108263 \ . \tag{3.69}$$

Once μ, a, J_2 are known, we can find an exact formula relating such quantities to e^2, namely

$$J_2 = \frac{e^2}{3}\left\{1 - \frac{4}{15}\frac{\omega^2 a^3}{\mu}\frac{e}{\left[\left(1+3\frac{1-e^2}{e^2}\right)\operatorname{arctg}\frac{e}{\sqrt{1-e^2}} - 3\frac{\sqrt{1-e^2}}{e}\right]}\right\} , \tag{3.70}$$

Despite its awful appearance, such a formula can be readily solved iteratively for e^2, once the value of the angular velocity of the Earth is fixed too, e.g. by astrogeodetic observations, namely

$$\omega = 7.292115 \cdot 10^{-5} \text{ rad s}^{-1} \ , \tag{3.71}$$

always referring to the GRS80 system. So now the eccentricity e of the ellipsoid can be computed too, from observable functionals of the actual gravity field. Once μ, a, e, ω are known, the value of U_0 is fixed by (3.67). U_0 is an important constant because it is used in the definition of the geoid. Just for the sake of completeness, let us recall as well that with the above constants one can also compute γ_0, finding the value reported in Table 3.1.

3.6 Definition of the Geoid

The geoid, $\mathcal{G}$, is that particular equipotential of the actual gravity field where the gravity potential W attains the value

$$W_0 = U_0 \ . \tag{3.72}$$

The ellipsoidal height of a point P on the geoid is called the geoid undulation, N_{P}; such a function $N_{\mathrm{P}} = N(\sigma)$ is used to represent its shape in ellipsoidal coordinates. We take as a known fact that all over the Earth

$$|N_{\mathrm{P}}| \leq 128 \text{ m, i.e. } 2 \cdot 10^{-5}\, a \ . \tag{3.73}$$

Traditionally the geoid was defined to coincide with the mean surface of the sea, that was believed, once time dependent phenomena like tides, waves, wind interaction, etc. were averaged on long time lags, to conform to an equipotential of the gravity field, due to an elementary hydrostatic reasoning.

Nowadays, having the possibility of directly observing the sea surface by satellite altimetry, it has been realized that, even subtracting the time variable component of the sea surface, the stationary surface left is significantly deviating from an equipotential surface. This is due to several factors, but one prominent among them is the presence in the ocean of steady streams, like the Gulf Stream, the Kuroshio, the circumpolar streams, just to mention a few; such currents in fact, due to the Coriolis force, generate small "mountains and valleys" on the sea surface. The vertical distance of this stationary sea surface from the geoid is called by definition the mean dynamic topography of the sea, η_{D}, so that the height of the stationary sea on the ellipsoid, h_{SS}, can be decomposed according to

$$\overline{h}_{\mathrm{SS}} = N + \overline{\eta}_{\mathrm{D}} \; ; \tag{3.74}$$

overbars in this formula are expressing long time averages. Assuming oceanographers to be able to properly model $\overline{\eta}_{\mathrm{D}}$, one realizes that (3.74) can be used to derive N on oceanic areas.

Apart from the game played by geodesists and oceanographers, similar to a dog biting its tail, we take as a fact, confirmed by physics and data, that when the geoid $\mathcal{G}$ is defined to pass close to tide gauges, $\overline{\eta}_{\mathrm{D}}$ is globally bounded to a few meters in magnitude, say

$$|\eta_{\mathrm{D}}| \le 2\ \mathrm{m}\,. \tag{3.75}$$

Now we are interested in analyzing how the different terms in (3.74) are changed when we move the value of a in a range of a few meters.

We aim first at proving that when a point $\mathrm{P}(\varphi)$ on the ellipsoid, with equatorial radius a, at latitude φ, is moved to $\mathrm{P}'(\varphi)$, a point on the ellipsoid with equatorial radius $a + \delta a$, at the same latitude φ, the shift $\delta \boldsymbol{r} = \boldsymbol{r}_{\mathrm{PP}'}$ is approximately given by

$$\delta \boldsymbol{r} \sim \delta a\, \boldsymbol{\nu}\,. \tag{3.76}$$

We use (2.74) rearranged in the approximate form

$$\begin{cases} \rho = a\left[1 + \dfrac{1}{2}e^2 \sin^2\varphi\right]\sin\varphi \\ z = a\left[1 - e^2\left(1 - \dfrac{1}{2}\sin^2\varphi\right)\right]\cos\varphi \end{cases} . \tag{3.77}$$

Remember that the quantities ω^2, μ and $Y = \mu\, J_2\, a^2$ are observable by Satellite Geodesy and so they are considered known and fixed. On the other hand, by using (3.70) to the first order in e^2, we can put $J_2 = \dfrac{e^2}{3}$ so that we have

$$Y = \frac{1}{3}\mu e^2 a^2 . \tag{3.78}$$

In (3.78) Y and μ are fixed; so, differentiating, we see that a change δa in a induces a change in e^2 according to

$$\delta\left(e^2\right) a = -2e^2\,\delta a . \tag{3.79}$$

Then going back to (3.77) and using (3.79), we get

$$\begin{aligned}
\delta\rho &= \delta a\left[1 + \frac{1}{2}e^2\sin^2\varphi\right]\sin\varphi + \frac{1}{2}\delta\left(e^2\right) a\,\sin^3\varphi = \\
&= \delta a\left[\sin\varphi - \frac{1}{2}e^2\sin^3\varphi\right] , \\
\delta z &= \delta a\left[1 - e^2\left(1 - \frac{1}{2}\sin^2\varphi\right)\right]\cos\varphi = \\
&= \delta a\left[\cos\varphi + e^2\left(1 - \frac{1}{2}\sin^2\varphi\right)\cos\varphi\right] .
\end{aligned}$$

Since

$$\left|\frac{1}{2}e^2\sin^3\varphi\right| < 3.5\cdot 10^{-3} , \quad \left|e^2\left(1 - \frac{1}{2}\sin^2\varphi\right)\cos\varphi\right| < 6.7\cdot 10^{-3} ,$$

we have, with a relative accuracy of 10^{-2},

$$\delta\boldsymbol{r} = \delta\rho\,\boldsymbol{e}_\rho + \delta z\,\boldsymbol{e}_z \cong \delta a\,\boldsymbol{\nu} . \tag{3.80}$$

We can observe that if we move by a height h along $\boldsymbol{\nu}$ we get a point that, leaving h unchanged as well as $\boldsymbol{\nu} = \boldsymbol{\nu}(a)$, undergoes the same shift as in (3.80), see Fig. 3.2.

We turn now to the value of U_0, by using (3.67) in the approximate form,

$$U_0 \cong \frac{\mu}{a}\left(1 - \frac{1}{2}e^2\right) + \frac{1}{3}\omega^2 a^2 .$$

Differentiating, using (3.79) and putting $\frac{\mu}{a^2} = \gamma_0$, we get

$$\begin{aligned}
\delta U_0 &= -\gamma_0\,\delta a\left(1 - \frac{1}{2}e^2\right) - \frac{1}{2}\gamma_0\,a\,\delta\left(e^2\right) + \frac{2}{3}\omega^2 a\,\delta a = \\
&= -\gamma_0\,\delta a\left(1 - \frac{3}{2}e^2 - \frac{2}{3}\frac{\omega^2 a}{\gamma_0}\right) .
\end{aligned}$$

Since

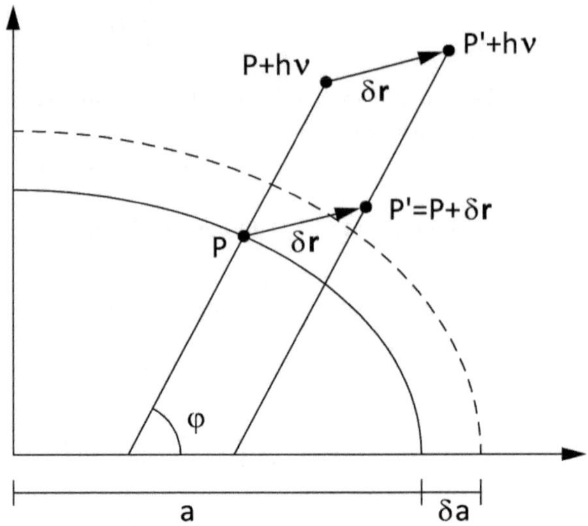

Fig. 3.2 The shift $\delta\boldsymbol{r}$ caused by a change of equatorial radius when φ, h are kept constant

$$\frac{3}{2}e^2 + \frac{2}{3}\frac{\omega^2 a}{\gamma_0} \sim 1.2 \cdot 10^{-2} ,$$

we see that at our level of accuracy

$$\delta U_0 \cong -\gamma_0\, \delta a . \tag{3.81}$$

When U_0 is changed to $U_0 + \delta U_0$, also the geoid will move to the equipotential surface with potential $W_0 + \delta W_0 = U_0 + \delta U_0$. The vertical shift L will then be

$$L = -\frac{\delta W_0}{g} .$$

On the other hand, $\boldsymbol{n}$ and $\boldsymbol{\nu}$ can be considered parallel at the level of 10^{-2} and g and γ_0 are such that

$$\frac{\gamma_0}{g} \sim 1 ,$$

always at the same level. Since clearly

$$\delta N = L - \delta a = -\frac{\delta W_0}{g} + \frac{\delta U_0}{\gamma_0} = -\frac{\delta U_0}{\gamma_0}\left(\frac{\gamma_0}{g} - 1\right) = \delta a \left(\frac{\gamma_0}{g} - 1\right) ,$$

we find $\delta N \cong 0$, with a relative error of the order of 10^{-2} with respect to δa.

Summarizing and returning to (3.74), we see that when we move a to $a + \delta a$ the point on the ellipsoid is raised δa, so $\overline{h}_{SS}$ will decrease by the same amount, because the stationary surface of the ocean is fixed in space. On the other hand N is not changed, so $\overline{\zeta}_{\mathrm{D}}$ will change by the same figure, δa.

Since a was a free parameter, at least at the level of a few meters, as discussed in the previous section, we see that now a can be used to minimize the mean square value of $\overline{\zeta}_{\mathrm{D}}$. This explains why, although irrelevant from the point of view of the approximation of W by U, a is still "estimated" up to centimeters as in Grafarend and Ardalan (1999), Burša et al. (2007).

Another useful remark is that the modern definition of geoid, namely (3.72), has to substitute the old practice of defining a "height" origin by using the equipotential surface passing through some tide gauge. This practice has indeed generated different equipotential surfaces as reference for various nations or group of nations. The unification of the different height datums, reconducting all of them to a unique geoid, i.e. to a unique reference for geodetic heights, is a problem known as "global height datum problem", which the geodetic community is facing nowadays. This will occupy us in the last chapter of the book.

References

Abramowitz M., Stegun I.A. (1964). *Handbook of mathematical functions with formulas, graphs, and mathematical tables*. Dover Publications, New York.

Burša M., Kenyon S., Kouba J., Šíma Z., Vatrt V., Vítek V., Vojtíšková M. (2007). *The geopotential value W_0 for specifying the relativistic atomic time scale and a global vertical reference system*. Journal of Geodesy, 81(2):103–110.

Grafarend E.W., Ardalan A.A. (1999). *World Geodetic Datum 2000*. Journal of Geodesy, 73(11):611–623.

Heiskanen W.A. and Moritz H. (1967). *Physical geodesy*. Freeman, San Francisco.

Hotine M. (1969). *Mathematical geodesy*. ESSA Monograph 2, U.S. Department of Commerce, Washington, DC.

Martinec Z. (1998). *Boundary value problems for gravimetric determination of a precise geoid*. LNES-Springer, Berlin.

Marussi A. (1985). *Intrinsic geodesy*. Springer, Berlin.

Moritz H. (1980). *Geodetic Reference System 1980*. Bulletin Gèodèsique, 62(3), 348–358.

Pizzetti P. (1894). *Sulla espressione della gravità alla superficie del geoide, supposto ellissoidico*. Atti della Reale Accademia dei Lincei, Rendiconti 3:166–172 (in Italian).

Sansò F., Sideris M.G. (2013). *Geoid determination: Theory and methods*. Lecture Notes in Earth System Sciences, Vol. 110. Springer-Verlag, Berlin, Heidelberg.

Somigliana C. (1929). *Teoria generale del campo gravitazionale dellellissoide di rotazione*. Memorie della Società Astronomia Italiana 4:541–599 (in Italian).

Somigliana C. (1930). *Sul campo gravitazionale esterno del geoide ellissoidico*. Atti della Reale Accademia dei Lincei, Rendiconti, 6:237–243 (in Italian).

Vanìcek P. and Krakiwsky E.J. (1986). *Geodesy: The concepts*, 2nd edn. Elsevier, Amsterdam.

Chapter 4
The Anomalous Potential and Its Determination

4.1 Outline

The knowledge of the normal potential and related ellipsoidal quantities are not enough to properly treat the problem of relating different types of geodetic heights.

To do that we need a much more precise knowledge of the geoid, i.e. of the gravity potential W, than that supplied by the ellipsoid, which leaves out the "last 100 m" of the geoid undulation. To do that we need to learn how to model the difference between W_{P} and U_{P}, namely the anomalous potential T_{P}. How this can be derived by a suitable fusion of different data sources, like surface gravity, satellite tracking, digital terrain models and oceanic mean dynamic heights, is certainly one of the main tasks of Physical Geodesy, requiring a good knowledge of some chapters of mathematics. We shall account here after of one of the main procedures along which the task is performed nowadays. We shall not go deeply into the mathematical background, but for the theorem of Runge-Krarup. The proof of the theorem, even in the simplified form provided here, needs not to be fully understood, however its consequences and implications need to be clearly visualized and kept in mind by the reader.

Although other approaches are present in geodetic literature, all of them need to go through two fundamental steps: the first is linearization of the relations expressing the observables as functionals of the potential, the second is to remove from our unknown T pieces that approximate its long wavelength behaviour as well as its short wavelength behaviour, controlled by the so called topographic signal. Such concepts are properly developed in the chapter. The rest is basically collocation theory as a technique to solve the relevant boundary value problem left for the residual part of T.

4.2 The Anomalous Potential

We define the anomalous potential T (P) as

$$T(\mathrm{P}) = W(\mathrm{P}) - U(\mathrm{P}) \ . \tag{4.1}$$

F. Sansò et al., *Geodetic Heights*, Springer Geophysics,
https://doi.org/10.1007/978-3-030-10454-2_4

Let us immediately observe that, since we have placed the polar axis of $\mathcal{E}$ along the rotation axis of the Earth, the centrifugal potential V_c (P) (see (3.9)) contained in both W (P) and U (P) is the same; therefore (see (3.12) and (3.58))

$$T\,(\mathrm{P}) = V_{\mathrm{N}}\,(\mathrm{P}) - V_{\mathrm{e}}\,(\mathrm{P}) \ . \tag{4.2}$$

Hence, since V_e (P) is harmonic outside $\mathcal{E}$ and even inside, for a depth of thousands of kilometers, from (4.2) and recalling (3.20) we find that T satisfies the Poisson equation

$$\Delta T\,(\mathrm{P}) = -4\pi G\rho\,(\mathrm{P}) \ ; \tag{4.3}$$

in particular T (P) *is harmonic outside the masses.*

Now let us remark as an empirical fact that, at the level of the topographic layer, the following relations of maximum order of magnitude hold

$$\left|\frac{T}{W}\right| \lesssim 2 \cdot 10^{-5} \ , \quad \frac{|g-\gamma|}{\gamma} \lesssim 10^{-4} \ . \tag{4.4}$$

This implies that T can be usefully considered as a quantity small of the first order, when we have to linearize functionals of W. However we have to underline that, if we try to go inside the masses, the behaviour of W and U (continued as a harmonic function) diverge one from the other (see Sansò and Sideris 2013), so that

$$\frac{|g-\gamma|}{\gamma} \lesssim 4 \cdot 10^{-3} \text{ at 20 km depth ,}$$
$$\frac{|g-\gamma|}{\gamma} \lesssim 2 \cdot 10^{-2} \text{ at 100 km depth .}$$

It follows that, some 20/30 km below the Earth surface, the significance of T (P) is lost and one should not use any more the actual normal potential to approximate W (P).

Having characterized the order of magnitude of T close to the masses, let us look now at its behaviour at infinity, i.e. for r tending to ∞. From (4.2) and recalling (3.19) and (3.68), one has

$$\begin{aligned} T\,(\mathrm{P}) &= W\,(\mathrm{P}) - U\,(\mathrm{P}) = V_{\mathrm{N}}\,(\mathrm{P}) - V_{\mathrm{e}}\,(\mathrm{P}) = \\ &= \left[\frac{\mu}{r} + \mathcal{O}\left(\frac{1}{r^3}\right)\right] - \left[\frac{\mu}{r} + \mathcal{O}\left(\frac{1}{r^3}\right)\right] = \mathcal{O}\left(\frac{1}{r^3}\right) \ . \end{aligned} \tag{4.5}$$

Notice that the above asymptotic relation comes from our choice to have the same value of $\mu = GM$ for the actual and normal potential, to put the origin at the barycentre of the masses, also coinciding with the centre of the ellipsoid $\mathcal{E}$, and to make the z axis coinciding with the rotation axis of the Earth as well as with the polar axis of $\mathcal{E}$.

A consequence of (4.5) is that, outside any Brillouin sphere of radius R, one can write the series expansion

$$T\left(\mathrm{P}\right)=\sum_{n=2}^{+\infty}\sum_{m=-n}^{n}T_{nm}\left(\frac{R}{r}\right)^{n+1}Y_{nm}\left(\sigma\right)\ . \tag{4.6}$$

Note that T_{nm} have the same dimensions as T, while in the literature, e.g. Sansò and Sideris (2013), we often find non dimensional T_{nm}^{nd}, related to the present coefficients by $T_{nm}^{\mathrm{nd}}=\left(\frac{\mu}{R}\right)^{-1}T_{nm}$. Here R is any radius close to the mean Earth radius.

In (4.6) the first two degrees, $\mathcal{O}\left(\frac{1}{r}\right)$ and $\mathcal{O}\left(\frac{1}{r^2}\right)$, are missing, complying with the asymptotic behaviour (4.5).

Let us recall as well here that, paralleling (3.54), the coefficients T_{nm} are functions of the chosen value for R because

$$T_{nm}=\frac{1}{4\pi}\int T\left(R,\sigma\right)Y_{nm}\left(\sigma\right)\mathrm{d}\sigma\ . \tag{4.7}$$

Now if we take any other sphere with radius $R'>R$, we have obviously

$$T\left(R',\sigma\right)=\sum_{n=2}^{+\infty}\sum_{m=-n}^{n}T_{nm}\left(\frac{R}{R'}\right)^{n+1}Y_{nm}\left(\sigma\right)\ ; \tag{4.8}$$

on the other hand, $T\left(R',\sigma\right)$ will have as well its own harmonic coefficients T'_{nm} such that

$$T\left(R',\sigma\right)=\sum_{n=2}^{+\infty}\sum_{m=-n}^{n}T'_{nm}Y_{nm}\left(\sigma\right)\ . \tag{4.9}$$

Comparing (4.8) and (4.9), one finds

$$T'_{nm}=\left(\frac{R}{R'}\right)^{n+1}T_{nm}\ . \tag{4.10}$$

Formula (4.10) represents the upward continuation of the harmonic coefficients of T from the sphere S_R to the sphere $S_{R'}$; as we can see, the upward continued coefficients T'_{nm} become exponentially smaller than the corresponding T_{nm} as the degree increases. This corresponds to a smoothing of T as a function of σ, going from $T\left(R,\sigma\right)$ to $T\left(R',\sigma\right)$.

However the converse is also true, namely if we start from the outer sphere $S_{R'}$ and its coefficients T'_{nm} and we want to derive the coefficients T_{nm}, i.e. the potential T, we will have an exponential increase of T'_{nm}, namely

$$T_{nm}=\left(\frac{R'}{R}\right)^{n+1}T'_{nm}\ . \tag{4.11}$$

So if we have an imperfect knowledge of T'_{nm}, namely

$$T'_{0nm} = T'_{nm} + \varepsilon_{nm} \ , \tag{4.12}$$

and we try to use the erroneous T'_{0nm} to derive $\widehat{T}_{nm}$ through (4.11), we get

$$\widehat{T}_{nm} = \left(\frac{R'}{R}\right)^{n+1} T'_{nm} + \left(\frac{R'}{R}\right)^{n+1} \varepsilon_{nm} = T_{nm} + \left(\frac{R'}{R}\right)^{n+1} \varepsilon_{nm} \ . \tag{4.13}$$

As we can see, $\widehat{T}_{nm}$ are equal to the true T_{nm} plus an error exponentially amplified. For instance, if ε_{nm} are just random errors, uncorrelated, with constant variance

$$\sigma^2 \left(\varepsilon_{nm}\right) = \sigma_\varepsilon^2 \ ,$$

as it happens if $T_0\left(R', \sigma\right)$ is equal to $T\left(R', \sigma\right)$ plus a white noise on the sphere $S_{R'}$, the error contaminating our estimate $\widehat{T}\left(R', \sigma\right)$ is

$$\delta T\left(R, \sigma\right) = \sum_{n=2}^{N} \sum_{m=-n}^{n} \left(\frac{R'}{R}\right)^{n+1} \varepsilon_{nm} Y_{nm}\left(\sigma\right) \ . \tag{4.14}$$

When the summation in (4.14) goes up to infinity, δT becomes an awkward random variable, with infinite variance, because, recalling (3.42)

$$\sum_{m=-n}^{n} Y_{nm}^2\left(\sigma\right) = (2n+1)\, P_n\left(1\right) = (2n+1) \ ,$$

we find

$$\sigma^2\left(\delta T\right) = \sum_{n=2}^{N} \left(\frac{R'}{R}\right)^{2n+2} (2n+1)\, \sigma_\varepsilon^2 \xrightarrow[N\to\infty]{} +\infty \ . \tag{4.15}$$

This shows that, if we try to make a downward continuation from the sphere $S_{R'}$ to the sphere S_R, we can expect a lot of fuzzy numbers because of the increasing variability of errors with the degree. In fact it is well known that, even assuming that we know exactly T'_{nm}, there are potentials that are harmonic outside $S_{R'}$ but not down to S_R, so that formula (4.11) cannot be meaningfully applied (see Moritz 1980, Sansò and Venuti 2010).

Note that the determination of T is an essential tool to be able to perform the transformation between several types of geodetic heights, so we have at least to be aware of how it is done, to handle the necessary calculations involving T.

The determination of T, starting from the historical approach of Stokes (1849), has always been done by building a model $\widehat{T}$ which is harmonic in a domain larger than Ω, i.e. harmonic even inside the masses down to some reference surface S_0, for instance an internal sphere S_{R_0} also called a Bjerhammar sphere. Since then $\widehat{T}$ seems to be a kind of downward continuation of T inside the masses, which is not, it is necessary to clarify the situation by illustrating a cornerstone of Physical Geodesy, namely the Runge-Krarup theorem.

4.3 The Runge-Krarup Theorem: A Mathematical Intermezzo

This is essentially a theorem saying that if we have a closed surface S, with Ω the exterior of S, and another internal surface S_0, with Ω_0 the exterior of S_0, such that $\overline{\Omega} \subset \Omega_0$, then any function harmonic in Ω can be approximated as well as we like by a function harmonic in Ω_0.

When we want to obtain a result of "approximation", we need to specify what this term means for us, i.e. we have to fix some topology for the space of functions harmonic in Ω. This can be done, as it was done by Krarup, in very general terms, but here we shall content ourselves to use the space mostly applied in geodetic literature, namely the space functions harmonic in Ω and such that their trace on S is in $L^2(S)$, i.e.

$$u \in \mathcal{H}(\Omega) \Rightarrow \Delta u = 0 \text{ in } \Omega\,, \quad \int_S u^2 \mathrm{d}S < +\infty\,. \tag{4.16}$$

Such a space is a Hilbert space with scalar product

$$\langle u, v\rangle_{\mathcal{H}} = \int u\, v\, \mathrm{d}S \tag{4.17}$$

and with the norm derived by (4.17). So $u_N \to u$, i.e. u_N approximates u as well as we like, in $\mathcal{H}$ means

$$\lim_{N\to\infty} \int (u_N - u)^2\, \mathrm{d}S = 0\,. \tag{4.18}$$

Similarly we can define the space $\mathcal{H}_0 = \mathcal{H}(\Omega_0)$ as

$$u_0 \in \mathcal{H}_0 \Rightarrow \Delta u_0 = 0 \text{ in } \Omega_0\,, \quad \int_{S_0} u_0^2\, \mathrm{d}S < +\infty\,. \tag{4.19}$$

We note that $\forall u_0 \in \mathcal{H}_0$ we can define a function $u_{0\Omega}$ which is the restriction of u_0 to $\overline{\Omega}$ (remember the $\overline{\Omega} \subset \Omega_0$), i.e. we can define a restriction operator $\mathcal{R}_\Omega : \mathcal{H}_0 \to \mathcal{H}$ such that

$$u_{0\Omega} = \mathcal{R}_\Omega u_0 \Rightarrow u_{0\Omega}(\mathrm{P}) \equiv u_0(\mathrm{P})\,, \quad \forall \mathrm{P} \in \Omega\,. \tag{4.20}$$

With the help of $\mathcal{R}_\Omega$ we can give the theorem a synthetic form.

Theorem 4.1 (Runge-Krarup)[1] *The set of functions*

$$U_0 \equiv \{\mathcal{R}_\Omega u_0 \,,\ u_0 \in \mathcal{H}_0\} \tag{4.21}$$

is densely embedded in $\mathcal{H}$.

This exactly means that $\forall u \in \mathcal{H}$ we can find $u_{0N} \in \mathcal{H}_0$ such that $u - \mathcal{R}_\Omega u_{0N} \to 0$ in $\mathcal{H}$. Since $\mathcal{H}$ is a Hilbert space, the above is equivalent to saying that there is no element $v \neq 0 \in \mathcal{H}$ which is orthogonal to U_0, i.e.

$$\forall u_0 \in \mathcal{H}\,,\ \langle v, \mathcal{R}_\Omega u_0\rangle_{\mathcal{H}} = 0 \ \Rightarrow\ v = 0\,. \tag{4.22}$$

We sketch here a proof without too many pretenses of rigorousness.
Take

$$u_0\,(\mathrm{P}) = \frac{1}{\ell_{\mathrm{PQ}}}\,,\quad \mathrm{Q} \in B_0 \text{ (interior of } S_0)\,;$$

it is obvious that $u_0 \in \mathcal{H}_0$, $\forall \mathrm{Q} \in B_0$. But then if $v \in \mathcal{H}$ is such that

$$V^v\,(\mathrm{Q}) = \langle v, \frac{1}{\ell_{\mathrm{PQ}}}\rangle_{\mathcal{H}} = \int_S \frac{v\,(\mathrm{P})}{\ell_{\mathrm{PQ}}}\,\mathrm{d}S_{\mathrm{P}} = 0 \qquad \forall \mathrm{Q} \in B_0\,,$$

we have that the single layer potential $V^v\,(\mathrm{Q})$ has to be zero in B_0. Since $V^v\,(\mathrm{Q})$ is harmonic in both B (interior of S) and Ω, and B_0 is an open set contained in B, $V^v\,(\mathrm{Q}) \equiv 0$ in B by the unique continuation property; namely, two functions u, v harmonic in some set B, that are equal in an open subset B_0 of B, have to coincide in the whole of B (Sansò and Sideris 2013). Indeed $V^v\,(\mathrm{Q})$ and 0 are precisely in the above situation.

On the other hand, imposing some regularity hypothesis on the surface S, it is known that a single layer with an $L^2\,(S)$ surface density is continuous throughout all of $\mathcal{R}^3$. This implies that $V^v\,(\mathrm{Q}) \equiv 0$ on S too. But then $V^v\,(\mathrm{Q})$ is harmonic in Ω, continuous in $\overline{\Omega}$ and zero on its boundary S, i.e. it has to be zero everywhere in Ω by the well known maximum principle, i.e. (see Sansò and Sideris 2013)

$$\max_{\mathrm{Q}\in\overline{\Omega}} V^v\,(\mathrm{Q}) = \max_{\mathrm{Q}\in S} V^v\,(\mathrm{Q})\,,\quad \min_{\mathrm{Q}\in\overline{\Omega}} V^v\,(\mathrm{Q}) = \min_{\mathrm{Q}\in S} V^v\,(\mathrm{Q})\,.$$

Since, as for any single layer (MacMillan 1958),

$$v\,(\mathrm{Q}) = \frac{1}{2\pi}\left\{\frac{\partial V^v\,(\mathrm{Q})}{\partial n_+} - \frac{\partial V^v\,(\mathrm{Q})}{\partial n_-}\right\}\,,$$

[1] *Note*: on historical ground Runge proved a similar theorem for analytic functions; the theorem was extended to harmonic functions by T. Krarup.

with $\boldsymbol{n}_{\pm}$ indicating the external/internal limit of the derivative along the normal to S, we find that $v \equiv 0$ and the theorem is proved.

Note that in the theorem S_0 is any closed surface with an open interior domain B_0. Now take a sequence S_k of such surfaces, internal one to the other and shrinking to some point that we take as the origin O, so that

$$B_k \supset \overline{B}_{k+1} \quad \text{or} \quad \Omega_{k+1} \subset \overline{\Omega}_k \,. \tag{4.23}$$

If we consider the corresponding Hilbert spaces $\mathcal{H}_k$, we have indeed

$$u_{k+1} \in \mathcal{H}_{k+1}\,, \quad \mathcal{R}_{\Omega_k} u_{k+1} \in \mathcal{H}_k\,, \quad \mathcal{R}_{\Omega}\mathcal{R}_{\Omega_k} u_{k+1} = \mathcal{R}_{\Omega} u_{k+1} \in \mathcal{H}\,,$$

so that

$$\mathcal{R}_{\Omega}\mathcal{H}_{k+1} \subset \mathcal{R}_{\Omega}\mathcal{H}_k \subset \cdots \subset \mathcal{H}\,, \tag{4.24}$$

each embedding being dense in $\mathcal{H}$. If we take the intersection

$$\bigcap_{k=0}^{+\infty} \mathcal{R}_{\Omega}\mathcal{H}_k = \mathcal{R}_{\Omega}\dot{\mathcal{H}}\,,$$

we get the restriction to Ω of all the functions that are harmonic outside the origin,

$$u \in \dot{\mathcal{H}} \ \Rightarrow \ \Delta u = 0 \qquad \forall \mathrm{P} \neq \mathrm{O}\,.$$

$\dot{\mathcal{H}}$ has not a Hilbert space structure, but this is not important to us. More interesting is that, if we take the sequence of solid spherical harmonics

$$S_{nm}(r, \sigma) = \left\{ \left(\frac{R}{r}\right)^{n+1} Y_{nm}(\sigma) \right\}\,,$$

we have indeed $S_{nm} \in \dot{\mathcal{H}}$ and so any finite linear combination of $\{S_{nm}\}$ is also in $\dot{\mathcal{H}}$, namely

$$u \in \mathcal{H}^F \equiv \left\{ \sum_{n=0}^{N} \sum_{m=-n}^{n} a_{nm} \left(\frac{R}{r}\right)^{n+1} Y_{nm}(\sigma) \right\} \Rightarrow u \in \dot{\mathcal{H}}\,.$$

In particular, what is of utmost importance for us is the following corollary of the Runge-Krarup theorem.

Corollary *The subspace $\mathcal{R}_{\Omega}\mathcal{H}^F$ is densely embedded into $\mathcal{H}(\Omega) \equiv \mathcal{H}$.*

This is rather obvious because writing the elements of $\mathcal{H}^F$ in the form

$$u \in \mathcal{H}^F \Rightarrow u = \sum_{n=0}^{N} \sum_{m=-n}^{n} a'_{nm} \left(\frac{R}{r}\right)^{n+1} Y_{nm}(\sigma)\,, \tag{4.25}$$

for some finite N, we see that, $\forall\varepsilon > 0$,

$$\mathcal{R}_{\Omega_\varepsilon}\mathcal{H}^F \subset \mathcal{H}(\Omega_\varepsilon) \qquad (4.26)$$
$$(\Omega_\varepsilon \equiv \{r \geq \varepsilon\}) \ ,$$

the embedding being dense, because $\left\{\left(\frac{R}{r}\right)^{n+1} Y_{nm}(\sigma)\right\}$ is an orthogonal, complete sequence in $\mathcal{H}(\Omega_\varepsilon)$. But then, for ε sufficiently small so that $\Omega_\varepsilon \supset \overline{\Omega}$, we get

$$\mathcal{R}_\Omega\mathcal{R}_{\Omega_\varepsilon}\mathcal{H}^F = \mathcal{R}_\Omega\mathcal{H}^F \subset \mathcal{R}_\Omega\mathcal{H}(\Omega_\varepsilon) \subset \mathcal{H} \ , \qquad (4.27)$$

each embedding being dense into the next.

Remark The neat result of the above mathematical discussion is that, given any potential $T \in \mathcal{H}(\Omega)$, we can find a $\widehat{T}_M \in \mathcal{H}^F$ that approximates T better than a prefixed level ε, or said in another way

$$\forall T \in \mathcal{H}(\Omega)\ , \ \forall\varepsilon > 0; \ \exists N_\varepsilon, \ \left\{\widehat{T}_{nm};\ n \leq N_\varepsilon\right\} \ \Rightarrow \ \left\|T - \widehat{T}_M\right\|_{\mathcal{H}} < \varepsilon$$

with

$$\widehat{T}_M = \sum_{n=2}^{N_\varepsilon}\sum_{m=-n}^{n} \widehat{T}_{nm}\left(\frac{R}{r}\right)^{n+1} Y_{nm}(\sigma) \ , \qquad (4.28)$$

for some fixed Bjerhammar radius R. A finite sum of the type (4.28) is called a global model of the anomalous potential.

It is clear from the above discussion that a global model $\widehat{T}_M$ is not a downward continuation of T; in addition there are many $\widehat{T}_M$ satisfying the same approximation level.

Actually we use validated models up to degree 2159 (e.g. EGM2008 Pavlis et al. 2012, 2013 or EIGEN-6C4 Förste et al. 2014, Shako et al. 2014), which have a resolution of about 10 km on the Earth surface. The use of these models however does imply calculations with about $4.6 \cdot 10^6$ coefficients T_{nm}, which is feasible but computationally heavy. Even more, if we wanted to reach the resolution of 1 km on S, we should use a model with 100 times coefficients than the above, what seems not particularly economical from the computational point of view.

So in our description on how to represent T, we shall always include a component of the type global model, but we shall leave to other methods a representation of the high resolution particulars of this potential.

4.4 Optimal Degree of Global Models, or Smoothing by Truncation

The decision to represent T by a global model $\widehat{T}_M$ leaves open the question of the degree N up to which $\widehat{T}_M$ should be developed and of which method should be employed to estimate the specific coefficients $\widehat{T}_{nm}$.

As for the second point, we could say that $\widehat{T}_{nm}$ are obtained by solving a specific boundary value problem, as it will be illustrated into the the next sections, while the first point will be discussed here.

In any event we assume that we have a tool that from some data is producing estimates

$$\widehat{T}_{nm} = T_{nm} + \varepsilon_{nm} \; , \tag{4.29}$$

where ε_{nm} are the estimation errors and Eq. (4.29) refers to some suitable radius R.

We call power of the degree n (or full power degree variances) the index

$$C_n\left(T\right) = \sum_{m=-n}^{n} T_{nm}^2 \tag{4.30}$$

and degree variances (we shall explain this term in the next section)

$$\sigma_n^2\left(T\right) = \frac{C_n\left(T\right)}{2n+1} \; . \tag{4.31}$$

Let us note that the quantity

$$\frac{1}{4\pi}\int T^2\left(R,\sigma\right) \mathrm{d}\sigma = \sum_{n=2}^{+\infty} C_n\left(T\right) < +\infty \tag{4.32}$$

has to be finite, implying that $C_n\left(T\right) \to 0$ for $n \to \infty$. Among others, this constitutes a necessary condition to be imposed on R. For instance, for EGM2008 R is close to be equal to a, the equatorial radius. Indeed we do not know exactly $C_n\left(T\right)$, but we can have a guess of them, $\widehat{C}_n\left(T\right) = C_n\left(\widehat{T}\right)$, by using $\widehat{T}_{nm}$. To be more precise, one could observe that the estimator $\widehat{C}_n\left(T\right) = \sum_{m=-n}^{n} \widehat{T}_{nm}^2$ is biased and $E\left\{\widehat{C}_n\left(T\right)\right\} = C_n\left(T\right) + \sum_{m=-n}^{n} \sigma^2\left(\varepsilon_{nm}\right)$; but then, if we assume to know $\sigma^2\left(\varepsilon_{nm}\right)$, we can easily construct the unbiased estimator $\overline{C}_n\left(T\right) = \widehat{C}_n\left(T\right) - \sum_{m=-n}^{n} \sigma^2\left(\varepsilon_{nm}\right)$.

It happens that by inspecting the plot of $\widehat{C}_n\left(T\right)$, for example computed from the EGM2008 coefficients (see Fig. 4.1), one can derive an empirical law for them (see

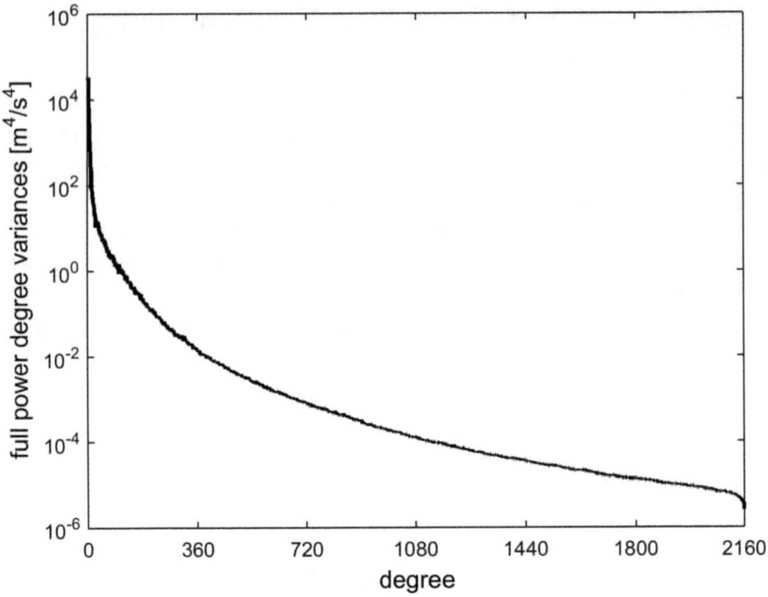

Fig. 4.1 The empirical full power degree variances of EGM2008

discussion in Sansó and Sideris 2013, Sect. 3.8). Older but evergreen models of C_n are also available, like those of Kaula (1966, 2000) and Tscherning and Rapp (1974).

Therefore we could say that, although we do not know the exact T_{nm}, we have a law for $C_n(T)$. This helps us to define the mean square omission error at degree N, i.e.

$$\mathcal{OE}_N^2 = \sum_{n=N+1}^{+\infty} C_n(T) \ ; \tag{4.33}$$

this is the error that we commit if instead of T we use just its development up to degree N. In fact if we split T (at the level of the sphere S_R) into

$$T = \sum_{n=2}^{N} \sum_{m=-n}^{n} T_{nm} Y_{nm}(\sigma) + \sum_{n=N+1}^{+\infty} \sum_{m=-n}^{n} T_{nm} Y_{nm}(\sigma) = T_{(N)} + T^{(N)} , \tag{4.34}$$

we see that, thanks to the orthogonality property of spherical harmonics, see (3.47),

$$\frac{1}{4\pi} \int \left[T - T_{(N)}\right]^2 d\sigma = \frac{1}{4\pi} \int \left[T^{(N)}\right]^2 d\sigma = \sum_{n=N+1}^{+\infty} C_n(T) = \mathcal{OE}_N^2 . \tag{4.35}$$

We note as well that $\mathcal{OE}_N$ is a decreasing function of N and it has to tend to 0 for $N \to \infty$, because of condition (4.32).

Now we observe that, recalling (4.29), one has

$$e\,(T) = T - \widehat{T}_M = T_{(N)} - \widehat{T}_M + T^{(N)} = \\ = \sum_{n=2}^{N} \sum_{m=-n}^{n} \varepsilon_{nm}\, Y_{nm}\,(\sigma) + \sum_{n=N+1}^{+\infty} T_{nm}\, Y_{nm}\,(\sigma)\ . \tag{4.36}$$

The mean square error of $e\,(T)$ over the unit sphere is then

$$\frac{1}{4\pi} \int e^2\,(T)\, \mathrm{d}\sigma = \sum_{n=2}^{N} \sum_{m=-n}^{n} \varepsilon_{nm}^2 + \mathcal{O}\mathcal{E}_N^2\ .$$

As we can see, this is still a random variable because it depends on ε_{nm}^2; so we can reasonably define a total error $\mathcal{E}_N^2$ as

$$\mathcal{E}_N^2 = E\left\{\frac{1}{4\pi} \int e^2\,(T)\, \mathrm{d}\sigma\right\} = \sum_{n=2}^{N} \sum_{m=-n}^{n} \sigma^2\,(\varepsilon_{nm}) + \mathcal{O}\mathcal{E}_N^2\ . \tag{4.37}$$

This is the total (mean square) error that we expect by substituting T with $\widehat{T}_M$. As we can see, it is in part due to the propagation of the estimation errors ε_{nm}, in part to the omission of the coefficients by truncating at degree N. The first term in (4.37) is called commission error

$$\mathcal{C}\mathcal{E}_N^2 = \sum_{n=2}^{N} \sum_{m=-n}^{n} \sigma^2\,(\varepsilon_{nm})\ . \tag{4.38}$$

As we said, it represents the effect of the estimation errors, up to degree N, which ultimately descend from the presence of measurement noise in the original data that have allowed to estimate the T_{nm} coefficients.

The terms

$$\sigma_n^2\,(\varepsilon) = \sum_{m=-n}^{n} \sigma^2\,(\varepsilon_{nm}) \tag{4.39}$$

are called error degree variances and we have

$$\mathcal{C}\mathcal{E}_N^2 = \sum_{n=2}^{N} \sigma_n^2\,(\varepsilon)\ . \tag{4.40}$$

As it is obvious, $\mathcal{C}\mathcal{E}_N^2$ is an increasing function of N and if for instance $\sigma^2\,(\varepsilon_{nm}) = \sigma_0^2$, as it happens when ε_{nm} are just white noise, then $\sigma_n^2\,(\varepsilon) = (2n+1)\,\sigma_0^2$ and indeed $\mathcal{C}\mathcal{E}_N^2 \to \infty$ when $N \to \infty$. This however is not the general case.

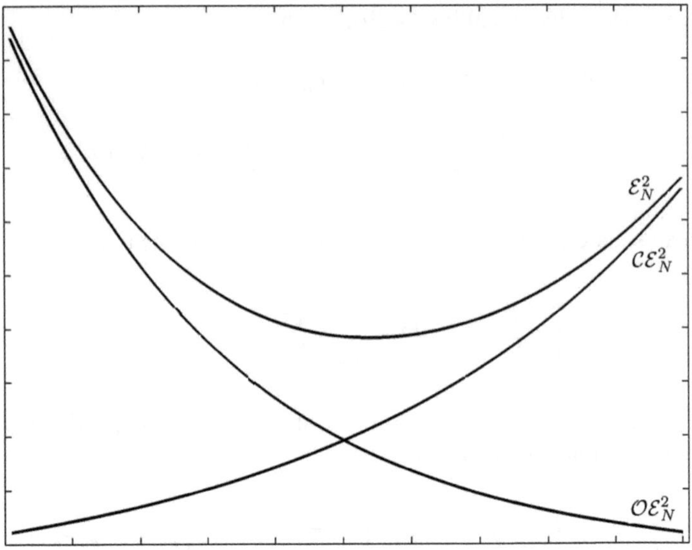

Fig. 4.2 The typical shape of $\mathcal{OE}_N^2$, $\mathcal{CE}_N^2$ and $\mathcal{E}_N^2$

Yet $\mathcal{E}_N^2$ as the sum of $\mathcal{CE}_N^2$ and $\mathcal{OE}_N^2$ will have a typical behaviour as shown in see Fig. 4.2, namely $\mathcal{E}_N^2$ will have a minimum at the degree $\overline{N}$ where the commission and omission errors cross. $\overline{N}$ is indeed our optimal choice for N, because the total error is minimum at this degree.

We note that the above condition implies

$$\sigma_{\overline{N}}^2(\varepsilon) = C_{\overline{N}}(T) \ ; \tag{4.41}$$

for instance, if $\sigma^2(\varepsilon_{nm}) = \sigma_0^2$, then $\sigma_{\overline{N}}^2(\varepsilon) = \left(2\overline{N}+1\right)\sigma_0^2$ and the optimal criterion is

$$\sigma_0^2 = \frac{C_{\overline{N}}(T)}{\left(2\overline{N}+1\right)} = \sigma_{\overline{N}}^2(T) \ .$$

This solves the posed problem. As a realistic example in Fig. 4.3 let us display the plot of potential error degree variances of a satellite model, when both the estimate of $\widehat{T}_{nm}$ is unregularized and it is conditioned by using $C_n(T)$ (see next section). As we can see, the optimal $\overline{N}$ in this case is around $\overline{N} = 250$.

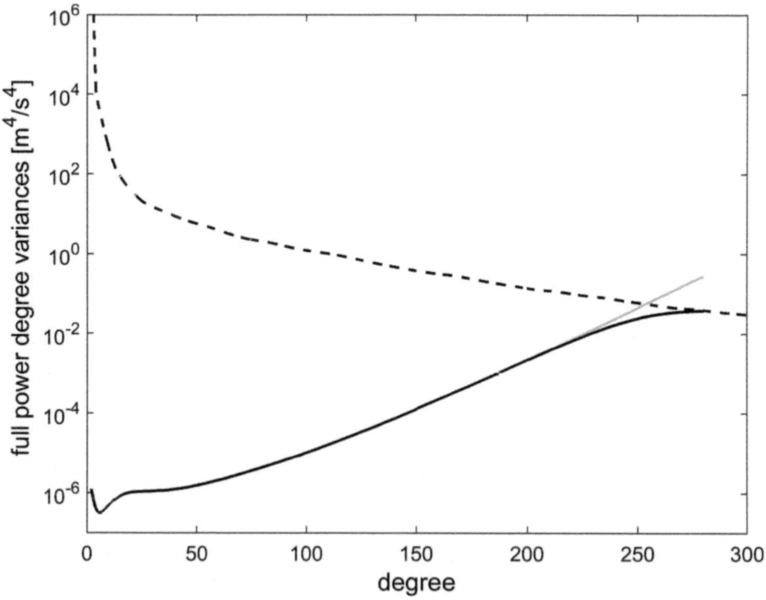

Fig. 4.3 Estimated error degree variances of a model from the ESA-GOCE mission with and without regularization, respectively in black and gray Brockmann et al. (2014), [Brockmann, personal communication, 2015]. The dash line shows the EGM2008 degree variances

4.5 Collocation Theory, or Smoothing by Prior Information

As in the previous section, we assume to know $T_{0nm} = T_{nm} + \varepsilon_{nm}$ up to some degree N, as well as the full power degree variances (4.30) and $\sigma_n^2(T)$. For the moment let us assume further on that ε_{nm} are independent from one another.

We want to state a criterion to estimate $\widehat{T}_{nm}$ that exploits, beyond the "observations" $\{T_{0nm}\}$, also the prior knowledge given by (4.30). In collocation theory this is done by establishing the minimum principle

$$\{\widehat{T}_{nm}\} = \text{ArgMin}\left\{\sum_{n=2}^{N}\sum_{m=-n}^{n}\frac{\left(T_{0nm}-\widehat{T}_{nm}\right)^2}{\sigma^2(\varepsilon_{nm})} + \sum_{n=2}^{N}\sum_{m=-n}^{n}\frac{\widehat{T}^2_{nm}}{\sigma_n^2(T)}\right\} . \tag{4.42}$$

As we can see, this is composed by a first quadratic functional that is essentially the same sum of squares as in least squares theory, while the second part of the functional has the purpose of stabilizing the solution as in Tikhonov theory. We observe also that this second functional would be the natural extension of least squares if we interpreted the prior information in terms of pseudo-observation equations

$$\begin{aligned} T_{0nm} &= 0 + \eta_{nm} \qquad \forall m\,,\ n > N \\ E\{\eta_{nm}\} &= 0\,,\ \ \sigma^2(\eta_{nm}) = \sigma_n^2(T)\ . \end{aligned}$$

This is also typical of a Bayesian interpretation in which every variable is stochastic by assumption.

All in all the principle (4.42) has an obvious, but significant, solution

$$\begin{cases} \widehat{T}_{nm} = \dfrac{\sigma_n^2(T)}{\sigma_n^2(T)+\sigma^2(\varepsilon_{nm})} T_{0nm} & (\forall m\,,\ n \le N) \\ \widehat{T}_{nm} = 0 & (\forall m\,,\ n > N) \end{cases} . \tag{4.43}$$

As we can see, the analogy to the Wiener-Kolmogorov filter is very strong (Sansó and Sideris 2013, Sect. 5.4).

Also here we truncate the estimated model at degree N, because there is no interaction between $\widehat{T}_{nm}$ $(n > N)$ and the observations. The coefficients of degree $n \le N$ are rescaled and not just equal to T_{0nm}. In particular at low degrees where we expect $\sigma_n^2(T) \gg \sigma^2(\varepsilon_{nm})$, we have $\widehat{T}_{nm} \sim T_{0nm}$, while for high degrees, where $\sigma_n^2(T) \to 0$ and $\sigma^2(\varepsilon_{nm})$ might even tend to a constant or in any way is expected to go zero much slower than $\sigma_n^2(T)$, we have that $\widehat{T}_{nm} \to 0$ much faster than T_{0nm}.

Remark There are significant examples in which T_{0nm} are directly derived from space observations. In these cases a stochastic model with independent estimation errors is too unrealistic; on the contrary the ε_{nm} have a fully populated covariance matrix C_ε. So if we reorganize T_{nm} in a vector $\boldsymbol{T}$ with some ordering and we introduce the diagonal matrix

$$K = \operatorname{diag}\left\{\sigma_n^2(T)\right\} ,$$

meaning that $\sigma^2(T_{nm}) = \sigma_n^2(T)$, $(m = -n, \ldots, 0, \ldots, n)$, the principle (4.42) is extended to

$$\min\left\{\widehat{\boldsymbol{T}}^T K^{-1}\widehat{\boldsymbol{T}} + (\boldsymbol{T}_0 - \widehat{\boldsymbol{T}})^T C_\varepsilon^{-1}(\boldsymbol{T}_0 - \widehat{\boldsymbol{T}})\right\} . \tag{4.44}$$

The variation equation of (4.44) is

$$\left(K^{-1} + C_\varepsilon^{-1}\right)\widehat{\boldsymbol{T}} = C_\varepsilon^{-1}\boldsymbol{T}_0$$

and its solution is given by

$$\widehat{\boldsymbol{T}} = K(K + C_\varepsilon)^{-1}\boldsymbol{T}_0 .$$

Such a formula is particularly nice because we do not need to invert two times the large matrix C_ε.

Anyway, what we have done up to now is basically to show how to filter a global model, where the coefficients themselves are considered as observations. On the other hand, we need a more general tool to treat the estimation of $\widehat{T}$ from a general set of observations; this is particularly important because the main sources of information on T come from gravity measurements and not from coefficients.

So to generalize the above discussion, we assume now to have a set of observations

$$m_{0i} = M_i\left(T\right) + \eta_i \qquad i = 1, 2, \ldots, M\ , \tag{4.45}$$

where $M_i\left(T\right)$ are linear functionals of T, namely numbers that linearly depend on T. We shall see in the next section how to write $M_i\left(T\right)$ for the main observables available.

We want to directly estimate $\widehat{T}\left(\mathrm{P}\right)$ at any point P in the harmonicity domain of $\widehat{T}$, recalling that by using the Runge-Krarup theorem $\widehat{T}$ is taken as harmonic down to a Bjerhammar sphere,

$$\begin{aligned}\widehat{T} &= \frac{\mu}{R}\sum_{n=2}^{+\infty}\sum_{m=-n}^{n}\widehat{T}_{nm}\left(\frac{R}{r}\right)^{n+1}Y_{nm}\left(\sigma\right) = \\ &= \frac{\mu}{R}\sum_{n=2}^{+\infty}\sum_{m=-n}^{n}\widehat{T}_{nm}S_{nm}\left(r,\sigma\right)\ .\end{aligned}$$

The new optimization principle then becomes

$$\min\left\{\sum_{i=1}^{M}\frac{\left[m_{0i} - M_i\left(\widehat{T}\right)\right]^2}{\sigma_{\eta_i}^2} + \sum_{n=2}^{+\infty}\sum_{m=-n}^{n}\frac{\widehat{T}_{nm}^2}{\sigma_n^2\left(T\right)}\right\}\ . \tag{4.46}$$

Leaving the proofs e.g. to the text Sansó and Sideris (2013, Sect. 5.5), we directly report here the solution of the principle (4.46). This can be obtained in terms of the so called covariance functions, hereafter defined

$$C\left(\mathrm{P},\mathrm{Q}\right) = \sum_{n=2}^{+\infty}\sum_{m=-n}^{n}\sigma_n^2\left(T\right)S_{nm}\left(r_{\mathrm{P}},\sigma_{\mathrm{P}}\right)S_{nm}\left(r_{\mathrm{Q}},\sigma_{\mathrm{Q}}\right)\ , \tag{4.47}$$

$$C\left(\mathrm{P},M_i\right) = \sum_{n=2}^{+\infty}\sum_{m=-n}^{n}\sigma_n^2\left(T\right)S_{nm}\left(r_{\mathrm{P}},\sigma_{\mathrm{P}}\right)M_i\left(S_{nm}\left(r_{\mathrm{Q}},\sigma_{\mathrm{Q}}\right)\right)\ , \tag{4.48}$$

$$C\left(M_k,M_i\right) = \sum_{n=2}^{+\infty}\sum_{m=-n}^{n}\sigma_n^2\left(T\right)M_k\left(S_{nm}\left(r_{\mathrm{P}},\sigma_{\mathrm{P}}\right)\right)M_i\left(S_{nm}\left(r_{\mathrm{Q}},\sigma_{\mathrm{Q}}\right)\right)\ . \tag{4.49}$$

The optimal solution is then obtained by the formula

$$\widehat{T}\left(\mathrm{P}\right) = \sum_{i,k=1}^{M}C\left(\mathrm{P},M_i\right)\left\{C\left(M_i,M_k\right) + \sigma_{\eta_i}^2\delta_{ik}\right\}^{(-1)}m_{0k}\ . \tag{4.50}$$

An important feature of the theory is that one can also compute the variance of the estimation error of $\widehat{T}\left(\mathrm{P}\right)$, namely

$$\mathcal{E}^2 (\mathrm{P}) = C (\mathrm{P}, \mathrm{P}) - \sum_{i,k=1}^{M} C (\mathrm{P}, M_i) \left\{ C (M_i, M_k) + \sigma_{\eta_i}^2 \delta_{ik} \right\}^{(-1)} C (\mathrm{P}, M_k) \ . \quad (4.51)$$

We cannot go here into the intricacy of the full estimation process and of its numerical implementation. However we shall make some comment on the use of (4.50) in a local area and on the remove-restore principle.

Remark (Collocation in a local refinement environment)

Assume we have global data sets, like satellite tracking or satellite gravity missions or just gravity observations all over the surface S; assume that we have solved the problem of estimating a global model $\widehat{T}_M$ from such global data sets, with a resolution regulated by its maximum degree. Now we have more observations, written as in (4.45), concentrated in a local area and we want to improve our knowledge of T in that area.

As a first operation we can remove the global information putting

$$m_{0i} = M_i (T_M + \delta T) + \eta_i = M_i (T_M) + M_i (\delta T) + \eta_i \ , \quad (4.52)$$

computing $M_i (T_M)$ and removing it from m_{0i}. We are left now with the unknown δT that represents the local behaviour of T. Before estimating δT with a formula like (4.50), it is usually convenient to further smooth the data by exploiting the information coming from a local Digital Terrain Model (DTM). In fact the fine variations of the topography produce a quite significant potential with an important content of high frequency. This is done by what is called the residual terrain correction and its potential δT_{tc}. In fact in general we have a much higher resolution in the knowledge of the topography than for other gravity measurements. This correction is called residual because we know that the long wavelength effect of topography is already captured by the model T_M. So in δT_{tc} we have to put the effect of the masses between the actual terrain and a smoothed version of it. This is usually done by discretizing the masses in prisms (Fig. 4.4).

So we now rewrite (4.52) as

$$\delta m_{0i} = m_{0i} - M_i (T_M) = M_i (\delta T_{\mathrm{tc}}) + M_i \left(\overline{\delta T} \right) + \eta_i \ , \quad (4.53)$$

where $M_i (\delta T_{\mathrm{tc}})$ is computed and removed from the known term (Sansò and Sideris 2013, Sect. 4.4). We are finally left with an unknown $\overline{\delta T}$, where long and short wavelengths have been removed or de-potentiated. It is now to $\overline{\delta T}$ that a collocation solution is applied. At the end we restore all the terms and $\widehat{T}$ is estimated in the area where we have added new measurements by the formula

$$\widehat{T} = \widehat{T}_M + \delta T_{\mathrm{tc}} + \overline{\delta T} \ . \quad (4.54)$$

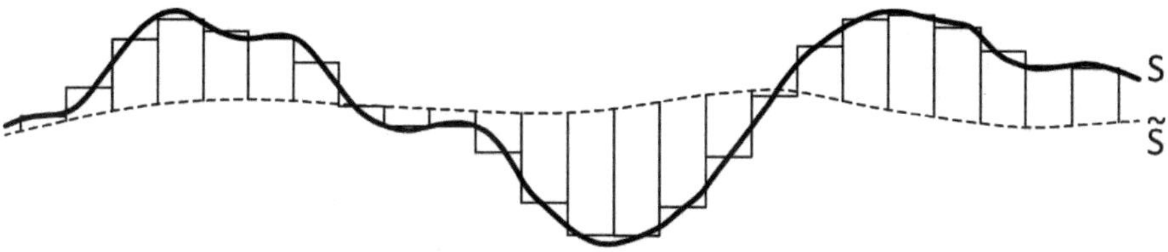

Fig. 4.4 Geometry of the terrain correction, i.e. potential generated by prisms including the masses between the actual topography S and a smoothed topography $\widetilde{S}$

4.6 On the Relation Between Potential and the Surface Gravimetric Observables

Let us remark first of all that W (P), and whence T (P), is related to several spatial observables that we shall not discuss in the present context, because this would require to enter into subjects of satellite dynamics that are far away from the main purpose of the book.

We shall mention however that due to the structure of satellite observation equations and the significant smoothing of T (P) at satellite altitudes, it comes natural that the processing of spatial geodesy data gives as an output the estimate of the harmonic coefficients $\{T_{nm}\}$ of T up to some maximum degree N. At present, with the data of the CHAMP, GRACE and GOCE missions, N can be taken to be as high as $N = 300$.

Given that, we come to the main observables on the surface of the Earth, that provide the major information on the gravity field.

4.6.1 Gravimetry

In principle gravimetry, in its absolute version, provides the measurement of the modulus of the gravity vector on continental areas.

In addition to absolute gravimeters, we have relative gravimeters that can observe the difference of gravity between two points. The old spring gravimeter schematized in Fig. 4.5a is nowadays substituted by superconducting gravimeters, see Fig. 4.5b, which are in principle able of measuring g with an accuracy of 1 μGal (10^{-6} Gal). Such relative measurements are arranged in networks constituted by closed loops, which, thanks to their redundancy, allow to estimate various biases in the observations.

All in all, also correcting the time variable part of g, we end up with a set of points $\{P_i\}$ (gravity stations), where we know

$$g\,(\text{P}) = |\boldsymbol{g}\,(\text{P})| = |\gamma\,(\text{P}) + \nabla T\,(\text{P})|\,. \tag{4.55}$$

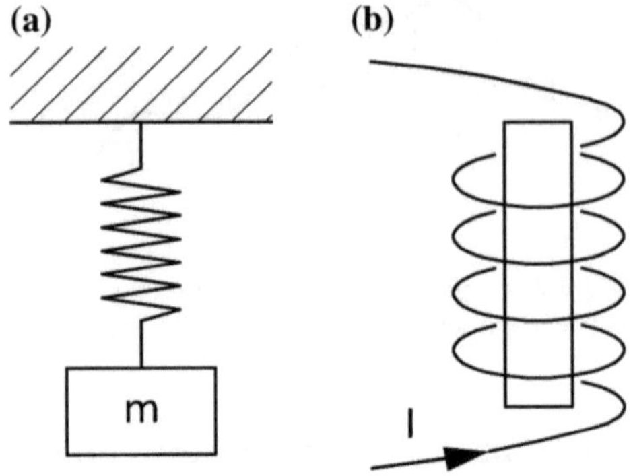

Fig. 4.5 The principle of measurement of gravity, **a** spring gravimeter (not in use any more), **b** superconducting gravimeter

The final accuracy with which we know $g\,(\mathrm{P})$ can be deemed to be somewhere between 0.1 and 0.01 mGal, which is certainly suitable for geodetic purposes. The actual data set at Bureau Gravimétrique International (BGI) comprises some 10^6 data, on continental areas, with a significant variability of points density. In particular South America, Africa and Antarctica are rather poorly covered by gravity observations.

Let us note explicitly that although nowadays gravity measurements are accompanied by the 3D ellipsoidal coordinates of P, given with sufficient accuracy by GPS observations, this is not the case for the largest part of the data existing in the BGI archives, where P_i have known horizontal coordinates but unknown ellipsoidal height h.

This imposes a particular manipulation of the equations, during linearization, which is characteristic of Physical Geodesy. We only mention that beyond continental gravity measurements, we have a marine gravity data set of direct gravity observations. This however is much less dense than the first and its accuracy is much lower (between 1 and 5 mGal). Furthermore, on oceans we have the more important data set of radar altimetry that we shall discuss hereafter.

Finally we have as well gravity data from aerogravimetry, in part on land and in part on sea; however it is only recently that such data have an accuracy below the mGal level and in any way we can think that they have been processed to provide grids of gravity values on the surface.

4.6.2 Levelling Combined with Gravimetry

Levelling is a kind of classical geodetic measurement, that is schematized in Fig. 4.6 for one of its constitutive steps. As we can see from the figure, the typical reading of a step of levelling is δL, which represents the projection of the vector $\boldsymbol{r}_{\mathrm{AB}} = \boldsymbol{r}_{\mathrm{B}} - \boldsymbol{r}_{\mathrm{A}}$ in the vertical direction $\boldsymbol{n}$ at the midpoint M between A and B, namely

$$\delta L = \boldsymbol{n} \cdot \boldsymbol{r}_{\mathrm{AB}} = -\frac{\boldsymbol{g}_{\mathrm{M}}}{g_{\mathrm{M}}} \cdot \boldsymbol{r}_{\mathrm{AB}} \; . \qquad (4.56)$$

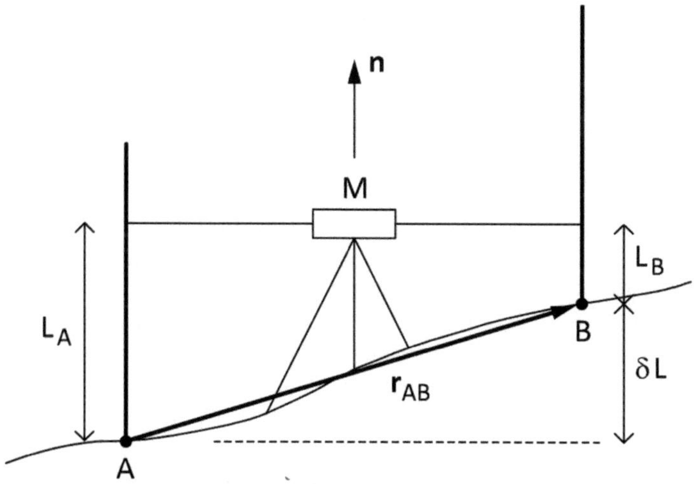

Fig. 4.6 The elementary operation of levelling with two vertical rods and two readings L_A, L_B from the middle in horizontal directions; the observation is $\delta L = L_A - L_B$

Such a measurement can be combined with the value of gravity at M, providing

$$g_M \delta L = -\boldsymbol{g}_M \cdot \boldsymbol{r}_{AB} = -\nabla W(M) \cdot \boldsymbol{r}_{AB} \cong W_A - W_B = -\delta W . \tag{4.57}$$

The last step in (4.57) is justified by the fact that in each levelling station the distance between A and B is typically between 100 and 200 m, so that $\boldsymbol{r}_{AB}$ can be considered to have an infinitesimal length, on a planetary scale.

We shall discuss in a dedicated chapter the levelling operations and their analytical formulation. Here we are interested in the fact that by adding the relation (4.57) along levelling lines, we can arrive to connect all the points of a certain region to an origin point P_0, which ideally could be placed on the geoid. This means that all over the surface of the continents we could arrive to know

$$W(P) = W_0 + \int_{P_0}^{P} dW = W_0 - \int_{P_0}^{P} g\, \delta L . \tag{4.58}$$

For several reasons, including the fact that it is difficult to state that P_0 is on the geoid, even if it is placed at a tide gauge, we could say that $W(P)$ is know, but for an additive constant. Even more such a constant is certainly different for different patches connected to different origins. So for the moment we shall overlook the problem of determining such constants, that will be treated in the last chapter of the book, and we shall assume that we know $W(P)$ at any point on land.

4.6.3 Radar-Altimetry on the Oceans

As already illustrated in Sect. 3.6, a radar-altimeter measures the height of a satellite on the ocean. The position of the radar-altimeter in space is known by GPS tracking at centimetric level; subtracting the former from the latter, we are left with the ellipsoidal height of the sea.

The footprint of the radar beam is regulated with a diameter between 100 and 1000 m, in such a way as to average the wave motion. Tides and barometric effects are modelled and subtracted from the observed height of the sea, so that by averaging in time we arrive at the (quasi) stationary sea surface. This one, in turn, is the sum of the geoid and the mean dynamic ocean topography η, which is related to geostrophic currents and provided by oceanographic models.

All together, one has the observation equation for H_0, i.e.

$$H_0 = h_\mathrm{R} - (N + \eta_t) + \nu \,, \tag{4.59}$$

with ν the observation noise, η_t the time dependent dynamic ocean topography, N the geoid undulation, h_R the ellipsoidal height of the radar-altimeter, see Fig. 4.7.

All the terms in (4.59), but for the unavoidable measurement error ν and the geoid N, are known or modelled. Hence (4.59) can be used to provide estimates of N over the ocean.

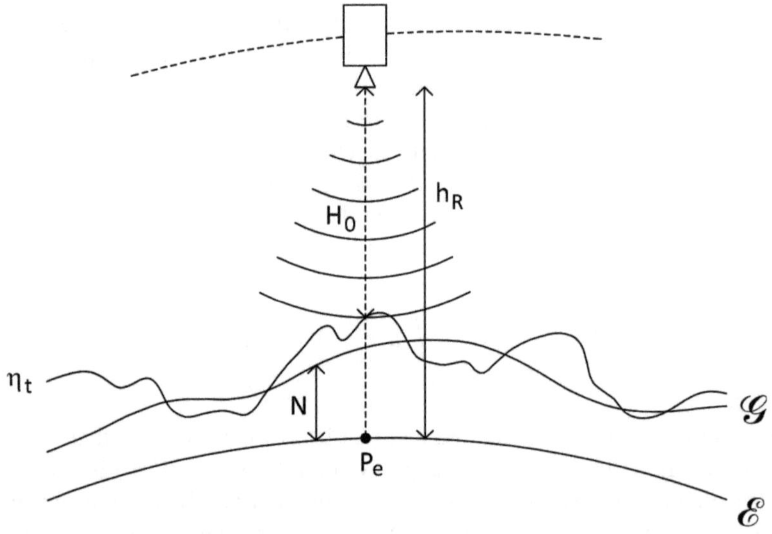

Fig. 4.7 Geometry of radar-altimetric observations: $\mathcal{E}$ ellipsoid, $\mathcal{G}$ geoid, N geoid undulation, η_t dynamic ocean topography, h_R ellipsoidal height of the radar-altimeter, H_0 radar-altimeter observation

Summarizing, and with a certain degree of abstraction, we could say that the main observables of Physical Geodesy can provide

$$\text{on continents: } W(\mathrm{P})\ ,\quad g(\mathrm{P}) \tag{4.60}$$

$$\text{on oceans: } \quad N(\mathrm{P_e})\ . \tag{4.61}$$

We recall that in (4.60) we know the horizontal coordinates (λ, φ) of P, but usually not its ellipsoidal height; in (4.61) $\mathrm{P_e}$ is on the ellipsoid and its (λ, φ) coordinates are known. However, note that from N one has also the third coordinate, namely (λ, φ, N) of the point $\mathrm{P}_{\mathcal{G}}$. So, recalling that on the geoid $W(\mathrm{P})$ has the known value W_0, one could substitute (4.61) with the relation

$$W(\mathrm{P}_{\mathcal{G}}) = W_0\ , \tag{4.62}$$

where $\mathrm{P}_{\mathcal{G}}$ has known ellipsoidal coordinates.

As a closing remark of the section, we recall again that further important sources of information on the gravity field are space geodetic methods, providing global models up to some degree N (nowadays we have $N \cong 300$), and digital terrain models, basically used at a local level to smooth the gravity field by residual terrain corrections.

To put together all this information is not an easy task; at a conceptual level, this is done by the so called Geodetic Boundary Value Problem theory that we shall review in the next section, especially with the purpose of providing the linearized version of the Eqs. (4.60) and (4.62), where the unknown field is not any more $W(\mathrm{P})$, but the anomalous potential $T(\mathrm{P})$.

4.7 The Geodetic Boundary Value Problem (GBVP)

In principle (4.60) and (4.62) can be put together, to formulate the following BVP:

to find $W(\mathrm{P}) = V(\mathrm{P}) + \frac{1}{2}\omega^2\rho^2$, with $V(\mathrm{P})$ regular harmonic in Ω, the exterior of surface S,

$$\begin{cases} \Delta V = 0 & \text{in } \Omega \\ V = \mathcal{O}\left(\dfrac{1}{r}\right) & r \to \infty \end{cases}\ ; \tag{4.63}$$

the surface S is composed by two patches, that we call L and O and correspond respectively to Land and Ocean,

$$S = L \cup O\ .$$

The surface O is geometrically known

$$\boldsymbol{r} \in O \ \Rightarrow\ \boldsymbol{r} = (\lambda, \varphi, N)\ ,\quad N = N(\lambda, \varphi) \tag{4.64}$$

and on O we know that the potential is constant, i.e.

$$W(\mathrm{P}) = W_0 \quad (\mathrm{P} \in O) \ ; \tag{4.65}$$

on the contrary, the surface L is unknown

$$\boldsymbol{r} \in L \ \Rightarrow \ \boldsymbol{r} = (\lambda, \varphi, h) \ , \quad h = h(\lambda, \varphi) \ (\text{unknown}) \tag{4.66}$$

but on L both the gravimetric quantities are known, i.e.

$$\begin{cases} W(\mathrm{P}) = W[\lambda, \varphi, h(\lambda, \varphi)] = W_0(\lambda, \varphi) \\ g(\mathrm{P}) = |\nabla W[\lambda, \varphi, h(\lambda, \varphi)]| = g_0(\lambda, \varphi) \end{cases} . \tag{4.67}$$

As such, this BVP can be classified as:

- a BVP for the Laplace operator in a space of regular harmonic functions (see (4.63)),
- a partially fixed boundary (see (4.64)) Dirichlet problem (see (4.65)),
- a partially free boundary (see (4.66)), mixed Dirichlet-Oblique Derivative (see (4.67)), because ∇W is not pointing towards the normal of S, non linear problem, because the second equation in (4.67) is highly non linear in the unknowns W and $h(\lambda, \varphi)$.

This is the GBVP in its most general form, or to be more precise, in its most general scalar form, as opposed to a vector form, previously stated in literature, where on L instead of knowing $(\lambda_\mathrm{P}, \varphi_\mathrm{P})$ it is considered as known the direction of $\boldsymbol{g}$ in an Earth-fixed reference frame. This vector form, though interesting, is certainly less realistic than the scalar one, because the data set of directions

$$\boldsymbol{n}(\mathrm{P}) = -\frac{\boldsymbol{g}(\mathrm{P})}{g(\mathrm{P})}$$

is essentially very poor and globally not very accurate. This is why we have chosen to directly present here the scalar GBVP. To the knowledge of the authors, this problem has never been rigorously analyzed in such a general formulation.

In any event, we shall go here to a linearization and a further simplification of the problem, conducting it to a form which is actually used to derive numerical solutions. We follow here the general approach introduced by Krarup (2006), although we want to mention as well Molodensky et al. (1962), Heiskanen and Moritz (1967) and Heck (1991). To this purpose, we notice that the problem has to be linearized with respect to all its unknowns, which here are the potential $W(\mathrm{P})$ as well as the height h_P of S corresponding to the land L. As for $W(\mathrm{P})$, it is only natural to put

$$W(\mathrm{P}) = U(\mathrm{P}) + T(\mathrm{P}) \ , \tag{4.68}$$

with $T(\mathrm{P})$ the variational unknown, and we shall put as well

$$h(\mathrm{P}) = \widetilde{h}(\mathrm{P}) + \zeta(\mathrm{P}) \ , \quad \mathrm{P} \in L \ , \tag{4.69}$$

where ζ (P), the variation of $\widetilde{h}$ (P), is called the generalized height anomaly; generalized because we shall reserve the name of proper height anomaly to a particular choice, that will be made in the sequel, for $\widetilde{h}$ (P).

In any way we recall that $\frac{T}{W} = \mathcal{O}\left(10^{-5}\right)$, so to keep in balance the linearization process we have also to put a constraint on $\widetilde{h}$, in such a way that $\frac{\zeta}{R} = \mathcal{O}\left(10^{-5}\right)$, with R the mean radius of the Earth, say 6371 km. This restricts the a-priori values of ζ (P) to be of the order of 100 m; such a choice is by the way consistent with the values of N (P), which are the counterparts of ζ (P) on the oceanic area.

We observe that the problem is indeed already linear for the Laplace equation in Ω, because

$$\Delta T = 0\,, \qquad \mathrm{P} \in \Omega\,;$$

however such a relation is of little use because Ω is not yet specified. In fact Ω has to be substituted by an approximate $\widetilde{\Omega}$, with a boundary $\widetilde{S}$ that includes $\left\{h = \widetilde{h}\right\}$ on L. For reasons that will be clearer later, instead of the actual known surface of O, we prefer in any way to make $\widetilde{S}$ to coincide with the ellipsoid $\mathcal{E}$ on the oceanic area. This is consistent with our previous discussion on orders of magnitude. In any way we notice that in doing so we modify the domain of harmonicity of the true T (P), yet, on account of the Runge-Krarup theorem, this does not prevent us from having an excellent approximation of T (P), neglecting only quadratic terms in the range $10^{-9} \div 10^{-10}$ of the potential. So we have an $\widetilde{\Omega}$ that is defined as the exterior of

$$\widetilde{S} \equiv \left\{h = \widetilde{h} \text{ on } L\,;\; h = 0 \text{ on } O\right\} \equiv \widetilde{S}_L \cup \widetilde{S}_O\,. \tag{4.70}$$

Naturally, to guarantee that $\widetilde{S}$ is a closed surface, one has to force $\widetilde{h}$ to go to zero on the coast lines. Therefore on $\widetilde{S}_O$ we can write

$$\begin{aligned} W\left(\mathrm{P_e}\right) &\cong W\left(\mathrm{P}\right) + g\left(\mathrm{P_e}\right) N \cong W\left(\mathrm{P}\right) + \gamma\left(\mathrm{P_e}\right) N = \\ &= W_0 + \gamma N \equiv U\left(\mathrm{P_e}\right) + T\left(\mathrm{P_e}\right) = U_0 + T\left(\mathrm{P_e}\right)\;, \end{aligned}$$

with $\mathrm{P_e} \in \widetilde{S}_O$, $\mathrm{P} \in O$. Recalling that $W_0 = U_0$, from the previous relation we derive the boundary condition for $\widetilde{S}_O$

$$T\left(\mathrm{P_e}\right) = \gamma\left(\mathrm{P_e}\right) N\left(\mathrm{P_e}\right) \qquad \mathrm{P_e} \in \widetilde{S}_O\,, \tag{4.71}$$

where the right hand side is known according to (4.64).

Coming to the land part $\widetilde{S}_L$, we have, considering the two points $\mathrm{P} \in S$ and $\widetilde{\mathrm{P}} \in \widetilde{S}_L$, along the same normal ν, at a distance ζ apart,

$$W\left(\mathrm{P}\right) = U\left(\mathrm{P}\right) + T\left(\mathrm{P}\right) \cong U\left(\widetilde{\mathrm{P}}\right) - \gamma\widetilde{\zeta} + T\left(\widetilde{\mathrm{P}}\right)\,. \tag{4.72}$$

Introducing the known potential anomaly

$$\mathrm{D}W = W\left(\mathrm{P}\right) - U\left(\widetilde{\mathrm{P}}\right)\,, \tag{4.73}$$

we write (4.72) as

$$\widetilde{\zeta} = \frac{T\left(\widetilde{\mathrm{P}}\right) - \mathrm{D}W}{\gamma}\ ; \tag{4.74}$$

this is known as the generalized Bruns relation. Notice that we use the non standard notation $\mathrm{D}W$ and (here below) $\mathrm{D}g$ to designate W and g anomalies and to distinguish them from ΔW and Δg that correspond to a particular choice of $\widetilde{S}$ and will be introduced later on.

Moreover, we have

$$\begin{aligned} g\,(\mathrm{P}) &= |\nabla U\,(\mathrm{P}) + \nabla T\,(\mathrm{P})| \cong \\ &\cong \gamma\,(\mathrm{P}) + \boldsymbol{e}_\gamma \cdot \nabla T\left(\widetilde{\mathrm{P}}\right) \cong \\ &\cong \gamma\left(\widetilde{\mathrm{P}}\right) + \gamma'\zeta + \boldsymbol{e}_\gamma \cdot \nabla T\left(\widetilde{\mathrm{P}}\right)\ ; \end{aligned} \tag{4.75}$$

here we have introduced the notation

$$\boldsymbol{e}_\gamma = \frac{\boldsymbol{\gamma}}{\gamma}\ ,\quad \gamma' = \frac{\partial \gamma}{\partial h}\ .$$

Considering that on $\widetilde{S}$ (see Sansò and Sideris 2013, Sect. 15.2)

$$\boldsymbol{e}_\gamma \cong -\boldsymbol{\nu}$$

with an accuracy of $5 \cdot 10^{-6}$, and introducing, similarly to (4.73), the gravity anomaly

$$\mathrm{D}g = g\,(\mathrm{P}) - \gamma\left(\widetilde{\mathrm{P}}\right)\ , \tag{4.76}$$

we can write (4.75) in the form

$$-\,\boldsymbol{\nu} \cdot \nabla T + \gamma'\widetilde{\zeta} \equiv -T' + \gamma'\widetilde{\zeta} = \mathrm{D}g\ . \tag{4.77}$$

Finally, using (4.74) in (4.77) and reordering, we get the fundamental equation of Physical Geodesy

$$-\,T'\left(\widetilde{\mathrm{P}}\right) + \frac{\gamma'}{\gamma} T\left(\widetilde{\mathrm{P}}\right) = \mathrm{D}g + \frac{\gamma'}{\gamma}\mathrm{D}W \qquad \widetilde{\mathrm{P}} \in \widetilde{S}_L\ . \tag{4.78}$$

Putting everything together, we find the linearized form of the scalar GBVP, namely

$$\begin{cases} \Delta T = 0 & \text{in } \widetilde{\Omega} \\ T = \gamma N & \text{on } \widetilde{S}_O \\ -T' + \dfrac{\gamma'}{\gamma} T = \mathrm{D}g + \dfrac{\gamma'}{\gamma}\mathrm{D}W & \text{on } \widetilde{S}_L \\ T = \mathcal{O}\left(\dfrac{1}{r}\right) & \end{cases}\ . \tag{4.79}$$

A first simplification of (4.79) is to fix explicitly the choice of $\widetilde{h}$. One possible useful choice, though not the only one, is to use the traditional condition

$$\mathrm{D}W = W\left(\mathrm{P}\right) - U\left(\widetilde{\mathrm{P}}\right) = 0 \,.$$

Such a condition gives $\widetilde{h}$ as the solution of the implicit function equation

$$U\left[\sigma, \widetilde{h}\left(\sigma\right)\right] = W\left(\sigma, h_{\sigma}\right) \;\Rightarrow\; \widetilde{h} = \widetilde{h}\left(\sigma\right) \,, \qquad \sigma = \left(\lambda, \varphi\right) \,. \tag{4.80}$$

With this choice, we shall denote

$$\widetilde{h} = h^{*} \,, \tag{4.81}$$

also called normal height, that we shall study in depth in the next chapter. Under such a choice, the corresponding

$$\zeta = h - h^{*} = \frac{T\left(\mathrm{P}^{*}\right)}{\gamma} \tag{4.82}$$

is the proper height anomaly and (4.82) is the proper Bruns relation. One can prove empirically that in fact $\mathcal{O}\left(|\zeta|\right) = 100$ m, which was one of the a-priori conditions to accept $\widetilde{h}$ as a suitable approximation of h.

We note as well that when P is on the geoid, as it happens in O, then

$$W\left(\sigma, h_{\sigma}\right) = W_0 = U_0 = U\left(\sigma, 0\right) \;;$$

in other words $h^{*} = 0$ and

$$\zeta_{\sigma} = N_{\sigma} \,, \quad \left(\mathrm{P} \in O\right) \,.$$

Another quantity that gets fixed by the choice (4.81) is the gravity anomaly that now is denoted as

$$\mathrm{D}g = g\left(\sigma, h_{\sigma}\right) - \gamma\left(\sigma, h^{*}_{\sigma}\right) \equiv \Delta g\left(\sigma\right) \,, \tag{4.83}$$

also called free air gravity anomaly. Notice that Δ in (4.83) has no relation with the Laplace operator.

The surface

$$S^{*} \equiv \left\{h = h^{*}\right\} = S^{*}_{L} \cup S^{*}_{O} \tag{4.84}$$

is called the Marussi telluroid (Marussi 1985); as we see, this is naturally a closed surface and this explains why we have chosen to use $S^{*}_{O} \subset \mathcal{E}$ as the approximate surface in the O region. In fact if we had chosen $S^{*}_{O} \equiv S_{O}$, which is possible because S_{O} is known, we would have for S^{*} a surface broken along the coast lines and this is not acceptable as boundary in a Boundary Value Problem.

In this way (4.79) becomes

$$\begin{cases} \Delta T = 0 & \text{in } \widetilde{\Omega} \\ T = \gamma N & \text{on } S_O^* \\ -T' + \dfrac{\gamma'}{\gamma} T = \Delta g & \text{on } S_L^* \\ T = \mathcal{O}\left(\dfrac{1}{r}\right) \end{cases} . \tag{4.85}$$

The solution of this problem is significantly complicated by the shape of the telluroid S_L^*, which mimics the geometry of the actual Earth surface in land areas, with irregular mountains as high as 10^{-3} R. In addition an important role is played by the geometry of the coasts, that separate S_O^* from S_L^*. So a further simplification is achieved by modifying the boundary condition on O, bringing it to the same form as that on L.

Without going into details, we only mention that, after a model up to some degree 200-300 is subtracted from T (see Rapp 1993), one goes locally from T (P) to Δg (P) by a slight generalization of the collocation theory outlined in Sect. 4.5.

More precisely when we subtract from T (P) a global model, e.g. up to degree 200, we theoretically obtain on O a signal containing only wavelengths below about 100 km. The covariance function of such a signal is decaying much faster than the original one and so a good prediction of Δg from T can be done in O even ignoring land data. By forming block averages, e.g. $5' \times 5'$ and using all available altimetric data, properly manipulated to eliminate biases (cross-over analysis), we finally arrive to determine a Δg field uniformly accurate at the level of about 2 mGal (see Sansò and Sideris 2013, Chaps. 6 and 9).

So the GBVP gets the form

$$\begin{cases} \Delta T = 0 & \text{in } \widetilde{\Omega} \\ -T' + \dfrac{\gamma'}{\gamma} T = \Delta g & \text{on } S^* \end{cases} ; \tag{4.86}$$

in (4.86) T is for the moment just a regular harmonic function in $\widetilde{\Omega}$.

Yet, with the new formulation we have introduced an important structural change into the problem. In fact, in contrast to (4.85), the solution of (4.86) is "almost" non unique. This can be better appreciated passing to the so called spherical approximation of (4.86), which consists in changing the boundary operator (but not the boundary S^*) into

$$-\frac{\partial}{\partial h} + \frac{\frac{\partial \gamma}{\partial h}}{\gamma} \cdot \cong -\frac{\partial}{\partial r} + \frac{\frac{\partial \gamma}{\partial r}}{\gamma} \cdot$$

and taking $\gamma = \frac{\mu}{r^2}$, so that (4.86) becomes

$$\begin{cases} \Delta T = 0 & \text{in } \widetilde{\Omega} \\ -\frac{\partial T}{\partial r} - \frac{2}{r} T = \Delta g & \text{on } S^* \end{cases} . \tag{4.87}$$

This is known as the simple Molodensky problem; would S^* be taken as a sphere, this becomes the Stokes problem, that we shall solve explicitly as an example below.

The theory of the simple Molodensky problem is contained in a few propositions (see Sansò and Sideris 2013, Sect. 15.4):

- first extend the definition of the (spherical approximation of) the gravity anomaly to the whole $\widetilde{\Omega}$, written in the form

$$-r \frac{\partial T}{\partial r} - 2T = r\,\Delta g_{\text{sph}} \equiv u\ ; \tag{4.88}$$

- verify by a direct computation that $r \frac{\partial T}{\partial r} = \boldsymbol{r} \cdot \nabla T$ is harmonic throughout $\widetilde{\Omega}$, so that $u = r\,\Delta g_{\text{sph}}$ is a harmonic function too in $\widetilde{\Omega}$;
- to derive the (regular) harmonic u in $\widetilde{\Omega}$, given its boundary values

$$u_0 = r\,\Delta g_{\text{sph}}\big|_{\widetilde{S}}\ , \tag{4.89}$$

is to solve the Dirichlet problem; this is very well known (see Sansò and Sideris 2013) to have a unique solution, for instance, if the boundary $\widetilde{S}$ is a Lipschitz surface (basically it admits conical points but not cusps) and $u_0 \in L^2\left(\widetilde{S}\right)$, i.e. $\int_{\widetilde{S}} u_0^2\,(\mathrm{P})\,\mathrm{d}S_{\mathrm{P}} < +\infty$;

- let $\overline{R}$ be any Brillouin radius, so that T and u are both harmonic in $\overline{\Omega} = \left\{r \geq \overline{R}\right\}$; let

$$\begin{aligned} T &= \sum_{n=0}^{+\infty} \sum_{m=-n}^{n} \overline{T}_{nm} \left(\frac{\overline{R}}{r}\right)^{n+1} Y_{nm}(\sigma) \\ u &= \sum_{n=0}^{+\infty} \sum_{m=-n}^{n} \overline{u}_{nm} \left(\frac{\overline{R}}{r}\right)^{n+1} Y_{nm}(\sigma) \end{aligned} , \tag{4.90}$$

then by a direct computation of (4.88) one finds the "spatial" relation

$$(n-1)\,\overline{T}_{nm} = \overline{u}_{nm}\ , \quad (|m| \leq n\ ;\ \ n = 0, 1, \ldots)\ ; \tag{4.91}$$

- Equation (4.91) implies that if u is derived from (4.88) then $\overline{u}_{1m} = 0$ $(m = -1, 0, 1)$, i.e. u has no terms of the type $\sum_{m=-1}^{1} c_{1m} \left(\frac{R}{r}\right)^2 Y_{1m}(\sigma)$ in its asymptotic expansion at infinity; we observe as well that if by any chance u is such that $\overline{u}_{00} = 0$, then we would have $\overline{T}_{00} = 0$ too, so that the asymptotic behaviour of T would be

$$T = \mathcal{O}\left(\frac{1}{r^3}\right), \tag{4.92}$$

as it was in our original definition of the anomalous potential;
- on the other hand, since

$$\delta T = \sum_{m=-1}^{1} c_{1m} \left(\frac{R}{r}\right)^2 Y_{1m}(\sigma) \tag{4.93}$$

is a function of r, homogeneous of degree -2, whatever are constants $c_{1\,-1}$, $c_{1\,0}$, $c_{1\,1}$, we see that δT is such that

$$r\,\delta T' + 2\,\delta T \equiv 0\,;$$

since δT is also obviously harmonic, outside the origin, we have that δT represents a null space of our BVP (4.87); this means that in any way a component like δT of T will never be fixed by the data;
- since in the end we want to find a solution T satisfying the traditional relation (4.92), we decide that the arbitrary δT should be fixed by the condition

$$\delta T \equiv 0\,,$$

that we know to be equivalent to placing the barycentre of T at the origin (or better placing the barycentre of U so as to coincide with that of W); furthermore we shall make some operation on the data $u_0 = r\,\Delta g_{\mathrm{sph}}\big|_{\tilde{S}}$, so that $\overline{u}_{00} = 0$ implying also that $\overline{T}_{00} = 0$, i.e. (4.92) holds true (see Sansò and Sideris 2013);
- if we do not want to put restrictions directly on u_0, we can change it by introducing four unknown constants, namely substituting the boundary condition $u|_S = u_0$ with

$$u|_S = u_0 + a\,\frac{R}{r}\bigg|_S + \sum_{m=-1}^{1} b_{1m} \left(\frac{R}{r}\right)^2 Y_{1m}(\sigma)\bigg|_S \tag{4.94}$$

and determining a, b_{1m} $(m = -1, 0, 1)$ in such a way that

$$\begin{aligned} &\overline{u}_{00} = 0 \text{ (to imply } \overline{T}_{00} = 0) \\ &\overline{u}_{1m} = 0 \text{ (to produce a boundary function } u_0 \\ &\qquad\qquad \text{that is } r \text{ times a spherical gravity anomaly);} \end{aligned}$$

one can prove that such conditions can always be satisfied by suitable constants $\forall u_0 \in L^2(S)$ (see Sansò and Sideris 2013);

- finally we derive $T = T(r, \sigma)$ by integrating radially (4.88) and taking into account that $u = \mathcal{O}\left(\frac{1}{r^3}\right)$, so that the closed expression is found

$$T(r, \sigma) = \frac{1}{r^2} \int_r^{+\infty} s\, u(s, \sigma)\, ds \; ; \tag{4.95}$$

one can directly prove that such a T satisfies (4.88), that it is a harmonic function and that it satisfies (4.92).

Summarizing, we have recalled the line showing that the simple Molodensky problem, modified as

$$\begin{cases} \Delta T = 0 & \text{in } \widetilde{\Omega} \\ -r\,T' - 2\,T = u_0 - a\,\dfrac{\overline{R}}{r} - \displaystyle\sum_{m=-1}^{1} b_{1m} \left(\frac{\overline{R}}{r}\right)^2 Y_{1m}(\sigma) & \text{on } S^* \\ T = \mathcal{O}\left(\dfrac{1}{r^3}\right) \end{cases} , \tag{4.96}$$

has one and only one solution $\{T, a, b_{1\,-1}, b_{1\,0}, b_{1\,1}\}$ whatever is the known term $u_0 \in L^2(S^*)$, i.e. $\Delta g \in L^2(S^*)$.

Once this is achieved, one can return to the original problem (4.86), that now we rewrite as

$$\begin{cases} \Delta T = 0 & \text{in } \widetilde{\Omega} \\ T' + \dfrac{\gamma'}{\gamma}\,T = \Delta g + a\,\dfrac{\overline{R}}{r^2} + \displaystyle\sum_{m=-1}^{1} b_{1m} \left(\frac{\overline{R}^2}{r^3}\right) Y_{1m}(\sigma) & \text{on } S^* \\ T = \mathcal{O}\left(\dfrac{1}{r^3}\right) \end{cases} , \tag{4.97}$$

and prove, by a perturbative argument, that (4.97) has a unique solution; however we are now obliged to put constraints on the inclination of the normal to S^* with respect to the radial direction $\boldsymbol{e}_r$ to guarantee the convergence of the perturbative process. Yet, a satisfactory result is obtained if we admit to a-priori know a model up to a maximum degree N, so that we can reduce our solution u to satisfy the asymptotic relation

$$u = \mathcal{O}\left(\frac{1}{r^{N+2}}\right) . \tag{4.98}$$

The theorem is the following (see Sansò and Sideris 2013, Sect. 15.4): if we know a model of T complete up to degree and order 20, then a unique solution to the Molodensky problem exists if the inclination of S^* with respect to $\boldsymbol{e}_r$ never exceeds 60°.

Fortunately, nowadays satellite geodesy is able, by analyzing data of low satellites, to provide the knowledge of the first 20 degrees of T with very high accuracy, in fact with an error of the order of 1 mm in terms of geoid. Such a knowledge has been pushed up to degree 200 with an error of about 2 cm, as we shall comment later on in Chap. 7.

As promised, we develop now the explicit solution of (4.96) when S^* is taken as a sphere, i.e. of the Stokes problem.

Example (Stokes theory)

Assume S^* is just a sphere with radius R_0; we want to solve the corresponding B.V.P. (4.96), which is of the simple Molodensky type.

Given our hypothesis, we expect T to be expandable into the spherical harmonic series

$$T = \sum_{n=0}^{+\infty} \sum_{m=-n}^{n} T_{nm} \left(\frac{R_0}{r} \right)^{n+1} Y_{nm}(\sigma) \ ; \tag{4.99}$$

this automatically satisfies the harmonicity condition. On the other hand we have, on the boundary,

$$u_0(\sigma) = R_0 \, \Delta g(\sigma) = R_0 \sum_{n=0}^{+\infty} \sum_{m=-n}^{n} \Delta g_{nm} Y_{nm}(\sigma) \ .$$

Since in this case we can take $\overline{R} = R_0$, we see that the known term in the second equation of (4.96) can be written as

$$\overline{u}_0 = R_0 \sum_{n=2}^{+\infty} \sum_{m=-n}^{n} \Delta g_{nm} Y_{nm}(\sigma) \ ,$$

if we make the choice

$$a = R_0 \, \Delta g_{00} \ , \qquad b_{1m} = R_0 \, \Delta g_{1m} \ ,$$

so that $u_{00} = u_{1m} \equiv 0$, $(m = -1, 0, 1)$. But in this case we know that $T_{00} = 0$ and, also, we can choose $T_{1m} = 0$ to satisfy the third equation of (4.96). Then for $n > 1$, we can use (4.91), i.e.

$$(n-1) \, T_{nm} = u_{nm} = R_0 \, \Delta g_{nm} \ .$$

Returning to the representation of T, we get

$$T\,(\mathrm{P}) = \sum_{n=2}^{+\infty} \sum_{m=-n}^{n} \frac{R_0}{n-1}\, \Delta g_{nm} \left(\frac{R_0}{r_\mathrm{P}}\right)^{n+1} Y_{nm}\,(\sigma_\mathrm{P})\ ;$$

Now we can remember that

$$\Delta g_{nm} = \frac{1}{4\pi} \int \Delta g\,(\sigma_\mathrm{Q})\, Y_{nm}\,(\sigma_\mathrm{Q})\, \mathrm{d}\sigma_\mathrm{Q}\ ,$$

so that the previous relation can be written as

$$T\,(\mathrm{P}) = \frac{1}{4\pi} \int \Delta g\,(\sigma_\mathrm{Q}) \left[\sum_{n=2}^{+\infty} \frac{R_0}{n-1} \left(\frac{R_0}{r_\mathrm{P}}\right)^{n+1} \sum_{m=-n}^{n} Y_{nm}\,(\sigma_\mathrm{P})\, Y_{nm}\,(\sigma_\mathrm{Q})\right] \mathrm{d}\sigma_\mathrm{Q} =$$

$$= \frac{R_0}{4\pi} \int \Delta g\,(\sigma_\mathrm{Q}) \sum_{n=2}^{+\infty} \frac{2n+1}{n-1}\, P_n\,(\cos\psi_\mathrm{PQ}) \left(\frac{R_0}{r_\mathrm{P}}\right)^{n+1} \mathrm{d}\sigma_\mathrm{Q}\ .$$

The series can be added in a closed form, obtaining the so called Stokes function (see Sansò and Sideris 2013, Sect. 3.4)

$$S\,(R_0, r_\mathrm{P}, \psi_\mathrm{PQ}) = \frac{2R_0}{\ell_\mathrm{PQ}} + \frac{R_0}{r_\mathrm{P}} - \frac{3R\,\ell_\mathrm{PQ}}{r_\mathrm{P}^2} - \frac{R^2}{r_\mathrm{P}^2}\cos\psi_\mathrm{PQ} \cdot$$

$$\cdot \left[5 + 3\log\frac{r_\mathrm{P} - R_0\cos\psi_\mathrm{PQ} + \ell_\mathrm{PQ}}{2r_\mathrm{P}}\right]\ ,$$

with

$$\ell_\mathrm{PQ} = \left[R_0^2 + r_\mathrm{P}^2 - 2R_0\, r_\mathrm{P}\cos\psi\right]^{\frac{1}{2}}\ .$$

So the solution of the Stokes problem is written in integral form as

$$T\,(\mathrm{P}) = \frac{R_0}{4\pi} \int S\,(R_0, r_\mathrm{P}, \psi_\mathrm{PQ})\, \Delta g\,(\sigma_\mathrm{Q})\, \mathrm{d}\sigma_\mathrm{Q}\ .$$

Let us remark that the GBVP theory, beyond providing a basis for the numerical determination of high degree anomalous models, is in itself one of the foundations of Physical Geodesy because it can specify what is the minimal information that can provide a stable solution T (P), under realistic conditions.

As claimed before, the solution of the GBVP is provided in terms of a finite sum of spherical harmonics of the type (4.99), truncated at a maximum degree N, which is called a global model of the anomalous potential. At present the most important of such models is EGM2008, which is complete up to degree and order 2159. The original data have been processed in such a way as to cover the Earth with a $5' \times 5'$ grid of area mean gravity anomalies; this corresponds to 9,331,200 values from which the model, described by 4,665,595 coefficients, is derived (see Sansò and Sideris 2013, Part II, Chap. 6). Another widely used global model, complete up to degree and order 2159, is EIGEN-6C4 that additionally includes GOCE data, though using the same EGM2008 $5' \times 5'$ grid of area mean gravity anomalies over continents. In 2020, it is foreseen the release of an updated version of EGM2008, called EGM2020, which will benefit from new data sources and procedures.

The overall error of the model, in terms of geoid, evaluated as a mean square estimation error over the whole Earth sphere, is considered to range around 5 cm; however the geographic distribution of the error, reflecting in particular areas of poor coverage of data and mountainous areas, shows that local error r.m.s. can amount up to 1 m.

The resolution of the model indeed cannot be better than the resolution of the input data, which in the average is around 10 km; this is reflected in the maximum degree 2159 chosen.

Indeed one might wonder whether, by using higher resolution data, one could improve the knowledge of the anomalous potential, at least locally. This is the case, although we cannot enter into details in this context; we rather send to literature, e.g. Sansò and Sideris (2013). Here we report only that an improved result can be obtained by first finely tuning the effect of local topographic masses on T (separately accounting for it) and then by applying a kind of local solver operator borrowed from random field prediction theory, for instance a collocation algorithm as recalled in Sect. 4.6, or some other equivalent techniques. What we fix here, about this more complex theory, is that there is a local solving operator S_A that acts on the improved data set $\{\Delta g\}$ in an area A, capable of producing a local anomalous potential

$$T = S_A(\Delta g) \tag{4.100}$$

that provides an approximation of the true T at the level of 2–3 cm in geoid, depending on the data available, the roughness of the surface (telluroid) and the roughness of the field Δg in A.

References

Brockmann J.M., Zehentner N., Höck E., Pail R., Loth I., Mayer-Gürr T., Schuh W.-D. (2014). *EGM TIM RL05: An independent geoid with centimeter accuracy purely based on the GOCE mission*. Geophysical Research Letters, 41(22):8089–8099.

Förste C., Bruinsma S.L., Abrikosov O., Lemoine J.M., Schaller T., Götze H.J., Ebbing J., Marty J.C., Flechtner F., Balmino G., Biancale R. (2014). *EIGEN-6C4: The latest combined global gravity field model including GOCE data up to degree and order 2190 of GFZ Potsdam and GRGS Toulouse*. In: 5th GOCE User Workshop, Paris. 25–28 November 2014. http://icgem.gfz-potsdam.de/ICGEM/documents/Foerste-et-al-EIGEN-6C4.pdf.

Heck B. (1991). *On the linearized Boundary Value Problem of physical geodesy*. Report No. 407, Department of Geodetic Science and Survey, The Ohio State University, Columbus.

Heiskanen W.A. and Moritz H. (1967). *Physical geodesy*. Freeman, San Francisco.

Kaula W.M. (1966). *Tests and combination of satellite determinations of the gravity eld with gravimetry*. Journal of Geophysical Research, 71:5303–5314.

Kaula W.M. (2000). *Theory of satellite geodesy*. Dover, New York.

Krarup T. (2006). *Letters on Molodenskiy's problem, III*. In: Borre K. (Ed.) Mathematical Foundation of Geodesy. Springer Verlag, Berlin, Heidelberg.

MacMillan W.D. (1958). *The theory of the potential*. In: Theoretical mechanics, Vol 2. Dover Publications, New York.

Marussi A. (1985). *Intrinsic geodesy*. Springer, Berlin.

Molodensky M.S., Eremeev V.F., Jurkina M.I. (1962). *Methods for study of the external gravitational field and figure of the Earth*. Translated from Russian, Israel Program for Scientific Translations, Jerusalem.

Moritz H. (1980). *Advanced physical geodesy*, 2nd edn. Wichmann, Karlsruhe.

Pavlis N.K., Holmes S.A., Kenyon S.C., Factor J.K. (2012). *The development and evaluation of the Earth Gravitational Model 2008 (EGM2008)*. Journal of Geophysical Research: Solid Earth, 117(B4):B04406.

Pavlis N.K., Holmes S.A., Kenyon S.C., Factor J.K. (2013). *Correction to "The development and evaluation of the Earth Gravitational Model 2008 (EGM2008)"*. Journal of Geophysical Research: Solid Earth, 118(5):2633.

Rapp, R.H. (1993). *Use of altimeter data in estimating global gravity models*. In: Sansò F., Rummel R. (Eds.), Satellite Altimetry in Geodesy and Oceanography, Lecture Notes in Earth Sciences, 50. Springer-Verlag, Berlin, Heidelberg, pp. 374–417.

Sansò F., Sideris M.G. (2013). *Geoid determination: Theory and methods*. Lecture Notes in Earth System Sciences, Vol. 110. Springer-Verlag, Berlin, Heidelberg.

Sansò F., Venuti G. (2010). *The convergence problem of collocation solutions in the framework of the stochastic interpretation*. Journal of Geodesy, 85(1):51–63.

Shako R., Förste C., Abrykosov O., Bruinsma S., Marty J.-C., Lemoine J.-M., Flechtner F., Neumayer K.-H., Dahle C. (2014). *EIGEN-6C: A High-Resolution Global Gravity Combination Model Including GOCE Data*. In: Flechtner F., Sneeuw N., Schuh W.-D. (Eds.), Observation of the System Earth from Space - CHAMP, GRACE, GOCE and future missions, Advanced Technologies in Earth Sciences. Springer, Berlin, Heidelberg, pp. 155–161.

Stokes G.G. (1849). *On the variation of gravity on the surface of the Earth*. Transactions of the Cambridge Philosophical Society, 8:672–695.

Tscherning C.C. and Rapp R.H. (1974). *Closed covariance expressions for gravity anomalies, geoid undulations, and deections of the vertical implied by anomaly degree variances*. Report No. 208, Department of Geodetic Science and Surveying, The Ohio State University, Columbus.

Chapter 5
Geodetic Coordinate Systems

5.1 Outline

In this chapter we return to the concept of coordinates, with a particular focus on various types of geodetic heights used in the geodetic environment.

We could say that there is a hierarchy of coordinate systems that goes from the most natural or physical ones, based on quantities related to the gravity field, to those more geometric, for which the gravity field plays almost no role.

The latter group includes the Cartesian triad, which constitutes the Terrestrial Reference System, co-rotating, in the mean, with the body of the Earth, with z axis along the rotation axis and the origin at the barycentre. Another coordinate system that shares the same characteristics is the terrestrial ellipsoidal coordinate system, which is centered at the same origin, namely the barycentre, has the polar axis along the rotation axis, is co-rotating with the Earth at the same mean angular velocity ω as (x, y, z) and has shape and dimension depending, as discussed in Sect. 3.5, on global gravimetric quantities like μ and J_2. The geometric properties of such systems have been discussed in Sects. 2.3.1 and 2.3.3, including the transformations of one into the other, so they will not be re-discussed here. We shall rather concentrate on the most natural coordinates, like the Hotine-Marussi system, the Helmert system, the Molodensky system. A particular care will be put in studying the transformations of such systems into ellipsoidal coordinates.

To achieve this, in particular for the so called orthometric heights, we will need to continue the potential and the gravity into the layer of the topographic masses. This can be done only by making some hypotheses on the mass density distribution and by applying suitable regularizing rules. Fortunately when this is needed only for the anomalous potential or gravity anomalies, as it is in our case, the result does not depend much on the error of the density model, so that the method can provide sensible answers.

It is for this reason that we shall open the chapter with a section on the subject of the continuation of the gravity field inside the masses.

F. Sansò et al., *Geodetic Heights*, Springer Geophysics,
https://doi.org/10.1007/978-3-030-10454-2_5

5.2 On the Continuation of Gravity into the Topographic Layer

We have already introduced in Sect. 3.3 the lines of the vertical $\{L_v\}$ and we have recalled the relation (3.28) that we repeat here

$$\frac{dg}{d\ell} = -2\mathcal{C}g + 4\pi G\rho - 2\omega^2 . \tag{5.1}$$

We observe that, if we assume to know the mean curvature $\mathcal{C}(\mathrm{P})$ and the mass density $\rho(\mathrm{P})$, the Eq. (5.1) can be taken as an ordinary differential equation for g that could be integrated along L_v from a point P on the surface S, where we assume to know the value $g(\mathrm{P}) = g_\mathrm{P}$, down to P_0 on the geoid $\mathcal{G}$, so to provide the value g_Q at any point on L_v (see Fig. 5.1).

We shall make two basic assumptions that will allow us to pursue the above program. Namely, calling $\ell = \ell_\mathrm{Q}$ the curvilinear coordinate along L_v, with origin in P_0 and positive upward, we shall assume that

$$\mathcal{C}(\mathrm{Q}) = \frac{1}{R + \ell} , \tag{5.2}$$

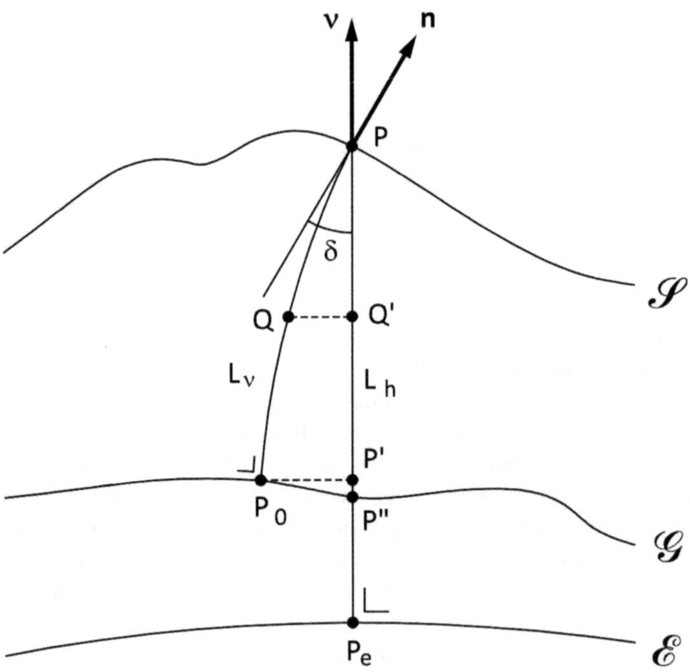

Fig. 5.1 The geometry of L_v and L_h (normal to $\mathcal{E}$) passing through the same point P, between the surface S and the ellipsoid $\mathcal{E}$

with R the mean radius of the Earth. A discussion of the error implied by such a drastic simplification can be found e.g. in Sansó and Sideris (2013, Chap. 2, page 92).

Moreover we shall assume that

$$\rho(\mathrm{Q}) = \rho_{\mathrm{C}} , \tag{5.3}$$

with $\rho_{\mathrm{C}} = 2.67$ g cm^{-3}, the credited mean value of the density of the crust. We shall see later on how small (or large) can be the error induced by (5.3).

If we call

$$q = 4\pi G \rho_{\mathrm{C}} , \quad p = q - 2\omega^2 \cong 0.22274 \text{ mGal m}^{-1}$$

the Eq. (5.1) can be integrated, giving

$$g(\mathrm{Q}) \equiv g(\ell) = \frac{(R+\ell_{\mathrm{P}})^2}{(R+\ell)^2} g_{\mathrm{P}} - \frac{p}{3}\left[\frac{(R+\ell_{\mathrm{P}})^3}{(R+\ell)^2} - (R+\ell)\right] , \tag{5.4}$$

where indeed $g_{\mathrm{P}} = g(\ell_{\mathrm{P}})$, ℓ_{P} are constants that we assume to know.

We note here that

$$\left(\frac{R+\ell_{\mathrm{P}}}{R+\ell}\right)^2 \geq 1 ,$$

indicating that in principle the integration could become unstable if we go very deep. In reality, for $\ell_{\mathrm{P}} \leq 6$ km, one has

$$\left(\frac{R+\ell_{\mathrm{P}}}{R+\ell}\right)^2 \leq \left(1+\frac{\ell_{\mathrm{P}}}{R}\right)^2 \cong 1 + 2 \cdot 10^{-3} ,$$

meaning that in the topographic layer the instability does not yet manifest itself sensibly. One has to remark as well that the hypothesis (5.2) has also a strong regularizing effect, because one can prove that $\mathcal{C}$ has a high frequency variability too, depending on the horizontal Laplacian of T. Yet the level of this effect can be considered as negligible in the present context, as commented in Sansó and Sideris (2013, Chap. 2, page 108). So we are left with the regularized downward continuation formula (5.4) for g.

Now, note that (5.4) can give us also the continuation of W to any point Q along L_v. In fact we have, by definition of L_v,

$$g = -\frac{\mathrm{d}W}{\mathrm{d}\ell}$$

so that

$$W(\mathrm{Q}) = W(\mathrm{P}) + \int_{\mathrm{Q}}^{\mathrm{P}} g \, \mathrm{d}\ell ,$$

the integral being computed along the vertical. With (5.4) and taking $Q \equiv P_0 \in \mathcal{G}$, so that $W(Q) = W_0$, we find

$$W_0 = W(P) + g_P \ell_P \left(1 + \frac{\ell_P}{R}\right) - \frac{p}{2}\ell_P^2 - \frac{p}{3}\frac{\ell_P^3}{R} . \tag{5.5}$$

A fast calculation of the order of magnitude of the last three terms in (5.5), divided by g_P to transform them in lengths and fixing $\ell_P \cong 6$ km, shows that the first term is of the order of ℓ_P, the second of the order of 3.6 m, the last of the order of 2.5 mm; this says that for all practical purposes, in the topographic layer, the last term can be safely neglected.

Finally let us assess the errors committed in continuing g down to the geoid ($\ell_Q = 0$) due to a model error in ρ. We assume that, as a maximum value, $\delta\rho = 10^{-1}\rho$. Then

$$\delta g = \frac{\delta p}{3} R\left[\left(1 + \frac{\ell_P}{R}\right)^3 - 1\right] \cong -\delta p\, \ell_P ,$$

with $\delta p \sim 2 \cdot 10^{-2}$ Gal km^{-1}. As we can see, in mountainous areas, where $\ell_P > 1$ km, the error in g can be very large, at least for such large errors in ρ. In any circumstance, in such areas we expect an error at least at the level of several mGal.

A further comment is that we assumed $g(P)$ and ℓ_P to be known; however, according to our discussion in Sect. 4.6, we can assume that $g(P)$ and $W(P)$ are known; so (5.5), where both $W(P)$ and W_0 are known, can be rather considered as a means to derive ℓ_P, neglecting as we said the last cubic term. In this case an error in ℓ_P is approximately given by

$$g_P\, \delta\ell - p\,\ell\,\delta\ell - \frac{\delta p}{2}\ell^2 = 0$$

or, with a justified simplification,

$$\delta\ell = \frac{\delta p\, \ell^2}{2 g_P} .$$

With $\delta p = 2 \cdot 10^{-2}$ Gal km^{-1} and $g_P \cong 10^3$ Gal, this gives

ℓ_P	1 km	2 km	4 km
$\delta\ell$	1 cm	4 cm	16 cm

showing that the error is small, but not completely negligible, especially for high mountains.

A last point has to be raised before closing the section, namely the Eq. (5.4) is significantly plagued by a systematic error, because it is derived from (5.1), which is exact, under the hypothesis (5.2), which is very rough. In interpreting (5.1) an error in $\mathcal{C}$ multiplies g, i.e. about 10^6 mGal; if we were able to transform (5.1) into an equation for a variational quantity like Δg, then the error in $\mathcal{C}$ would multiply something of the order of 10^2 mGal, reducing significantly its impact. This is possible indeed, as shown in Sansó and Sideris (2013, Part I, Sect. 2.4), because, similarly to (5.1), one can write the equation for the normal gravity

$$\frac{d\gamma}{d\ell} = -2\mathcal{C}_0\,\gamma - 2\omega^2\,, \tag{5.6}$$

where we have taken into account that the normal vertical lines have the same length as the ellipsoidal height (cf. (3.66)), so that $\ell \cong h$ in this case, and that $\rho = 0$ for the normal field outside $\mathcal{E}$. Then subtracting (5.6) from (5.1) and exploiting the appraisal

$$|\mathcal{C} - \mathcal{C}_0| \leq \frac{10^{-3}}{R}\,, \tag{5.7}$$

one arrives at the equation ($q = 4\pi G\rho$)

$$\frac{\partial \Delta g}{\partial \ell} = -2\mathcal{C}_0\,\Delta g + q\,, \tag{5.8}$$

where the hypothesis $\mathcal{C}_0 \sim \dfrac{1}{R+\ell}$ produces an error of the order of 10^{-7} mGal $\mathrm{m}^{-1} \ll q$.

5.3 The Hotine-Marussi Triad (Λ, Φ, W)

The two astrogeodetic coordinates $\Sigma = (\Lambda, \Phi)$, respectively longitude and latitude, are related to the direction of the vertical $\boldsymbol{n}$ and its Cartesian components in the geocentric (x, y, z), by the relation

$$\boldsymbol{n} = -\frac{\boldsymbol{g}}{g} = \begin{vmatrix} \cos\Phi\cos\Lambda \\ \cos\Phi\sin\Lambda \\ \sin\Phi \end{vmatrix}. \tag{5.9}$$

As $\boldsymbol{n}$, Σ can be determined by astrogeodetic observations, that first recover $\boldsymbol{n}$ in a celestial system and then rotate the vector to reckon its components in the terrestrial system (for details, see for instance Vanìcek and Krakiwsky 1986). W is just the gravity potential and it completes the triad.

As already observed in the introduction, the function $W(\mathrm{P})$ cannot be used as height throughout the whole exterior space. Indeed on the equatorial plane, at a distance of about seven times the radius of the Earth, W attains a minimum value and then it starts increasing for $r \to \infty$.

A different question is whether (Λ, Φ, W) can constitute a real, unambiguous coordinate system in our layer of interest. This is true if on the equipotential surfaces deployed in our relevant region, it never happens that $\boldsymbol{n}$ can become parallel at two different points; this in fact would mean that two different points have the same coordinates. This is the same to say that equipotential surfaces in the topographic layer are convex.

Although it is not impossible to find mass distributions that create non-convex equipotentials (Bocchio 1981), this seems not to be the case for the Earth. So we shall assume that in the topographic layer, (Λ, Φ, W) constitutes a coordinate system without singular points, at least in the correspondence $(\Lambda, \Phi, W) \rightarrow \mathrm{P}$; the inverse correspondence indeed displays the typical singularity of spherical coordinates already discussed in Sect. 2.3.

One important statement concerns the coordinate line L_w. This in fact is defined to be the line along which $\Lambda = \Lambda_0$, $\Phi = \Phi_0$, both being constant, i.e.

$$L_w \equiv \{\mathrm{P} \; ; \; \boldsymbol{n}(\mathrm{P}) = \boldsymbol{n}_0\} \; , \tag{5.10}$$

with $\boldsymbol{n}_0 = \boldsymbol{n}(\mathrm{P}_0)$ and P_0 is any point, e.g. on the geoid, on which $W = W_0$. The point is that, if $\boldsymbol{n}_0$ is orthogonal to the geoid, $\{W = W_0\}$ at P_0, the same is not any more true for points P on which $W(\mathrm{P}) < W_0$, because the equipotential surfaces $\{W = \overline{W} \; ; \; \overline{W} < W_0\}$ are not parallel to the geoid, as discussed in Sect. 3.3. The lines L_w are called, according to T. Krarup, isozenithal lines and, as we see, they are not coinciding with the lines of the vertical L_v. The situation is illustrated in Fig. 5.2.

It is possible to write the differential equation of isozenithal lines by the reasoning that we sketch hereafter.

We start from (5.9) and we note that, by an elementary differential calculus, when we move $\mathrm{P} \equiv \{\boldsymbol{r}\}$ by an infinitesimal $\mathrm{d}\boldsymbol{r}$ so that $\boldsymbol{g}(\boldsymbol{r})$ goes into $\boldsymbol{g}(\boldsymbol{r} + \mathrm{d}\boldsymbol{r})$, we have

$$\mathrm{d}\boldsymbol{n} = -\frac{1}{g}(I - P_n)\,\mathrm{d}\boldsymbol{g} \; , \tag{5.11}$$

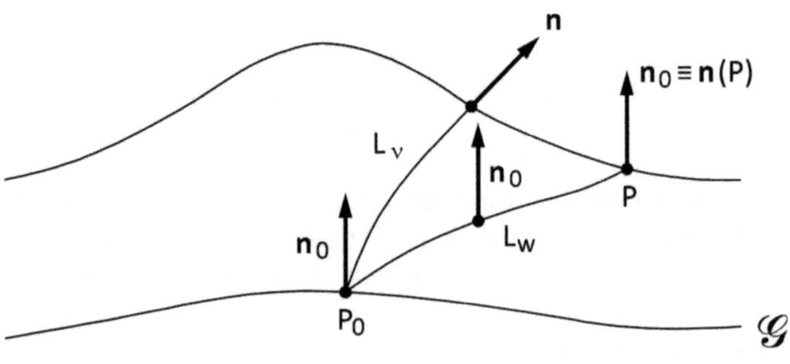

Fig. 5.2 The different paths of L_v and L_w through P_0

where $P_n \, \mathrm{d}\boldsymbol{g} = \boldsymbol{n} \, (\boldsymbol{n} \cdot \mathrm{d}\boldsymbol{g})$ is the projection of $\mathrm{d}\boldsymbol{g}$ on $\boldsymbol{n}$. Now, when $\mathrm{d}\boldsymbol{r}$ is along L_w, i.e. $\mathrm{d}\boldsymbol{r} = \boldsymbol{e}_w \, \mathrm{d}\ell$, we must have $\mathrm{d}\boldsymbol{n} = 0$. So along L_w

$$\mathrm{d}\boldsymbol{g} = P_n \, \mathrm{d}\boldsymbol{g} \tag{5.12}$$

has to hold. But (5.12) says that

$$\mathrm{d}\boldsymbol{g} \parallel \boldsymbol{n} \, . \tag{5.13}$$

On the other hand

$$\mathrm{d}\boldsymbol{g} = M \mathrm{d}\boldsymbol{r} = M \boldsymbol{e}_w \, \mathrm{d}\ell \, ,$$

with M the Marussi tensor (Marussi 1985), i.e. in Cartesian coordinates

$$M = \left[\frac{\partial^2 W}{\partial x_i \, \partial x_k} \right] .$$

So (5.13) says that

$$M \boldsymbol{e}_w \parallel \boldsymbol{n}$$

or, also recalling that $\boldsymbol{n} = \boldsymbol{n}_0$ along L_w,

$$\boldsymbol{e}_w \parallel M^{-1} \boldsymbol{n} \equiv M^{-1} \boldsymbol{n}_0 \, .$$

So finally the equation of L_w is determined by the tangent field

$$\boldsymbol{e}_w = \frac{\mathrm{d}\boldsymbol{r}}{\mathrm{d}\ell} = \frac{M^{-1} (\boldsymbol{r}) \, \boldsymbol{n}_0}{|M^{-1} (\boldsymbol{r}) \, \boldsymbol{n}_0|} \, .$$

It might be a nice exercise for the reader to verify that, with a purely spherical potential, $W_S = \frac{\mu}{r}$, one has $\boldsymbol{n}_0 = \boldsymbol{e}_{0r}$ and $M = \frac{\mu}{r^3} (I - 3P_r)$, so that $M^{-1} \div \left(I - \frac{3}{2} P_r \right)$ and we get then $\boldsymbol{e}_w = \boldsymbol{e}_r = \boldsymbol{e}_{0r}$; therefore, in this particular case, $L_v \equiv L_w$.

Remark 5.1 (Geopotential numbers and dynamic heights)

As it is obvious, W has the counterintuitive behaviour that it decreases when we move upward. This inconvenience can be eliminated by defining a *geopotential number* C as

$$C\,(\mathrm{P}) = W_0 - W\,(\mathrm{P}) \, , \tag{5.14}$$

because indeed $C\,(\mathrm{P})$ increases from lower to higher equipotential surfaces. We note that when $\mathrm{P} = \mathrm{P}_0$, a point on the geoid, then $W\,(\mathrm{P}_0) = W_0$ and $C\,(\mathrm{P}_0) = 0$.

We can observe as well that, if we had chosen the alternative definition of geoid as the equipotential surface passing through a given point P_0, e.g. a tide gauge station, then $C(\mathrm{P})$ would become observable by levelling and gravimetry (see Sect. 4.6), without knowing W_0. Since this is the practice adopted in many countries, we shall return in the last chapter to this point to explain how to unify the different datums.

Furthermore we can say that both $W(\mathrm{P})$ and $C(\mathrm{P})$ are dimensionally gravity potentials, namely the square of a velocity. To bring back a potential coordinate to the dimension of a length, as it seems intuitive for a height coordinate, sometimes a different coordinate is introduced, called *dynamic height* and defined by

$$H^{\mathrm{D}}(\mathrm{P}) = \frac{C(\mathrm{P})}{\overline{\gamma}_0} = \frac{W_0 - W(\mathrm{P})}{\overline{\gamma}_0} , \tag{5.15}$$

where $\overline{\gamma}_0$ is any *constant* value close to the actual gravity, i.e. to 10^3 Gal. As $\overline{\gamma}_0$, it could be convenient, for instance, to take the mean value of γ on the ellipsoid, namely

$$\overline{\gamma}_0 = 979.7614249 \text{ Gal} ,$$

although any other constant value, close by, would do. As we shall see later, H^{D} so defined results to be close to other types of heights, particularly to the orthometric heights.

It has to be remarked that in any way W, C, H^{D}, together with $\Sigma = (\Lambda, \Phi)$, share the same geometric behaviour, in particular in relation to isozenithal lines and coordinate surfaces.

Finally we have to understand how the Hotine-Marussi triad is related to geometrical coordinates. Since (x, y, z) and (λ, φ, h) can be just mathematically transformed one into the other, i.e. they are geometrically equivalent, we can study only the transformation between (Λ, Φ, W) and (λ, φ, h). The inverse transformation is obvious, in the sense that if we know $W(\mathrm{P}) = W(\lambda, \varphi, h)$, then $\mathrm{P} \rightarrow W$ is given and $\Sigma = (\Lambda, \Phi)$ is determined by inverting (5.9), namely

$$\left| \begin{array}{c} \cos\Phi \cos\Lambda \\ \cos\Phi \sin\Lambda \\ \sin\Phi \end{array} \right| = \frac{\nabla W(\mathrm{P})}{|\nabla W(\mathrm{P})|} \tag{5.16}$$

or

$$\begin{aligned} \mathrm{tg}\Lambda &= -\frac{g_y}{g_x} \\ \mathrm{tg}\Phi &= -\frac{g_z}{g} . \end{aligned}$$

So turning to the direct transformation, we have to show how to pass from (Λ_P, Φ_P, W_P) to $(\lambda_P, \varphi_P, h_P)$. We continue to assume that $W(P)$ is a known function of the ellipsoidal coordinates of P and so $T(P)$ is known too.

We have already introduced in Sect. 4.7 the point P^*, which is characterized by the fact that it is on the same ellipsoidal normal as P, i.e.

$$\begin{cases} \lambda_{P^*} = \lambda_P \\ \varphi_{P^*} = \varphi_P \end{cases} \quad \text{or} \quad \boldsymbol{\nu}_{P^*} = \boldsymbol{\nu}_P$$

and it has an ellipsoidal height h^* such that (4.80) is satisfied, namely

$$U_{P^*} = W_P \ . \tag{5.17}$$

Now we introduce another point $P^{*\prime}$ (see Fig. 5.3), which is characterized by the conditions

$$\begin{cases} \lambda_{P^{*\prime}} = \Lambda_P \\ \varphi_{P^{*\prime}} = \Phi_P \end{cases} \quad \text{or} \quad \boldsymbol{\nu}_{P^{*\prime}} = \boldsymbol{n}_P \tag{5.18}$$

and

$$U_{P^{*\prime}} = W_P \ ; \tag{5.19}$$

as for (5.17), also (5.19) can be used to derive $h^{*\prime} = h(P^{*\prime})$. The couple of Eqs. (5.18) and (5.19) are known as Marussi mapping and by them the ellipsoidal coordinates of $P^{*\prime}$ are known.

The key point here is that the vector $\boldsymbol{r}_{P^{*\prime}P}$ is of the maximum order of

$$\mathcal{O}(|\boldsymbol{r}_{P^{*\prime}P}|) = 100 \text{ m} \tag{5.20}$$

and for such small vectors one can put

$$T(P) \cong T(P^{*\prime}) \ ; \tag{5.21}$$

in fact, if we take $\mathcal{O}(|\nabla T|) = 10^2$ mGal for a shift of 10^2 m, one has as a maximum $\dfrac{\delta T}{\gamma} \sim 1$ cm, i.e. $\delta T \sim 10^{-4}\, T$, which is acceptable.

First of all we notice that

$$h^{*\prime} \cong h^* \ ;$$

this is intuitive from Fig. 5.3 and it is confirmed by noting that P^* and $P^{*\prime}$ have the same value of U, so δh^* can be computed by differentiating the last formula in Table 3.2, truncated to the first order in h. Since U varies only with h and φ, the result is approximately

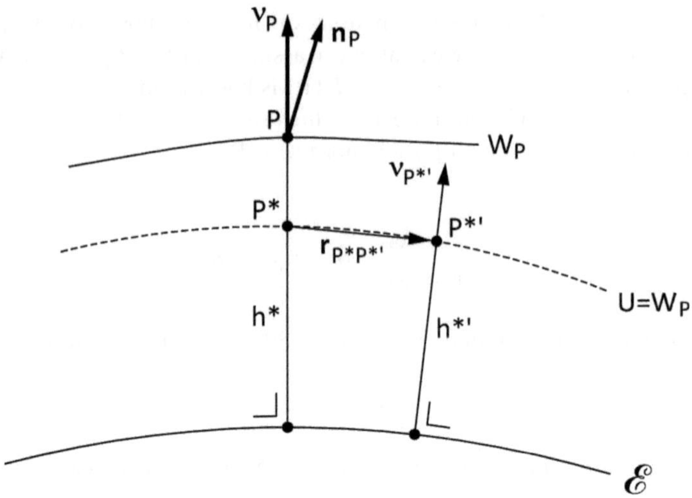

Fig. 5.3 The three points P, P*, P*'; $\mathcal{O}(\zeta) \sim \mathcal{O}(|\boldsymbol{r}_{P^*P^{*\prime}}|) \sim 10^2$ m; note that P*, P*' lay on the same equipotential of $U(P)$, so $\boldsymbol{r}_{P^*P^{*\prime}} \cdot \boldsymbol{\nu}_P \cong 0$

$$\delta h^* \cong \frac{-\gamma_{e\varphi}\, h\, \delta\varphi}{\gamma_e} ;$$

for a shift of $\delta\varphi \sim 2 \cdot 10^{-5}$, corresponding to $\sim$ 120 m, and an altitude of 6 km, this is below the μm and therefore we can forget it.

On the other hand we have as well

$$\zeta(P^{*\prime}) = \frac{T(P^{*\prime})}{\gamma(P^{*\prime})} ,$$

which is computable because we know the ellipsoidal coordinates of $P^{*\prime}$; but it is easy to verify that

$$\zeta(P^*) \cong \zeta(P^{*\prime}) ,$$

also taking (5.21) into account. Therefore we can put

$$h = h^* + \zeta(P^*) \cong h^{*\prime} + \zeta(P^{*\prime}) ,$$

so that h is now known.

Coming to the horizontal coordinates (λ_P, φ_P), we have first of all

$$\boldsymbol{n}_P = -\frac{\boldsymbol{\gamma}_P + \nabla T}{|\boldsymbol{\gamma}_P + \nabla T|} ; \tag{5.22}$$

but

$$|\gamma_P + \nabla T| \cong \gamma_P + \boldsymbol{e}_\gamma \cdot \nabla T = \gamma_P - \widetilde{\boldsymbol{n}} \cdot \nabla T ,$$

because $\boldsymbol{e}_\gamma = \dfrac{\boldsymbol{\gamma}_P}{\gamma_P} = -\widetilde{\boldsymbol{n}}_P$.

On the other hand $\mathcal{O}(|\widetilde{\boldsymbol{n}} - \boldsymbol{\nu}|) \sim 5 \cdot 10^{-6}$, which multiplied by ∇T goes down to the μGal level, so we can say

$$|\gamma_P + \nabla T| = \gamma_P - \boldsymbol{\nu} \cdot \nabla T$$

and therefore

$$|\gamma_P + \nabla T|^{-1} = \frac{1}{\gamma_P}\left(1 + \frac{\boldsymbol{\nu} \cdot \nabla T}{\gamma_P}\right) . \tag{5.23}$$

Substituting (5.23) in (5.22) and keeping only first order terms, we get

$$\boldsymbol{n}_P = \widetilde{\boldsymbol{n}}_P - \frac{1}{\gamma_P}\left[\nabla T - \boldsymbol{\nu} \cdot (\boldsymbol{\nu} \cdot \nabla T)\right] . \tag{5.24}$$

The vector

$$\boldsymbol{\delta} = \boldsymbol{n}_P - \boldsymbol{\nu}_P \tag{5.25}$$

is called the vector deflection of the vertical and its modulus $\delta = |\boldsymbol{\delta}|$ just deflection of the vertical, a quantity that being generally small (of the order of $3 \cdot 10^{-4}$ at most) is approximately equal to the angle between $\boldsymbol{n}_P$ and $\boldsymbol{\nu}_P$.

The vector

$$\widetilde{\boldsymbol{\delta}} = \widetilde{\boldsymbol{n}}_P - \boldsymbol{\nu}_P , \tag{5.26}$$

that we already encountered in scalar terms in (3.63), is the normal vector deflection of the vertical and we know that in the topographic layer $\mathcal{O}(\widetilde{\delta}) = \mathcal{O}(|\widetilde{\boldsymbol{\delta}}|) \sim 5 \cdot 10^{-6}$; more precisely we know that $\widetilde{\boldsymbol{\delta}}$ is pointing northward, in the northern hemisphere, so that

$$\widetilde{\boldsymbol{\delta}} \cong \frac{\gamma_{e\varphi}}{\gamma_0}\frac{h}{a}\boldsymbol{e}_\varphi \cong 5.3 \cdot 10^{-3} \sin 2\varphi \frac{h}{a}\boldsymbol{e}_\varphi . \tag{5.27}$$

It is immediate to verify that computing $\widetilde{\boldsymbol{\delta}}$ with $\varphi = \Phi$ and $h = h^*$ does not change significantly its value, so we consider it a known vector.

So returning to (5.24) we can write, subtracting $\boldsymbol{\nu}_P$ to both members,

$$\boldsymbol{\delta} = \widetilde{\boldsymbol{\delta}} - \frac{1}{\gamma}(I - P_\nu)\nabla T , \tag{5.28}$$

with P_ν the projector in the $\boldsymbol{\nu}$ direction. Note that we do not specify any more where the terms multiplying ∇T are computed, because we know that it makes no difference whether this is in P, P^* or $\mathrm{P}^{*\prime}$; so we shall assume that they are computed in $\mathrm{P}^{*\prime}$, which is known.

Finally, going back to the definition (5.25) and observing that

$$\boldsymbol{n}_\mathrm{P} = \boldsymbol{\nu}_{\mathrm{P}^{*\prime}} = \boldsymbol{\nu}\,(\Lambda, \Phi)$$

while

$$\boldsymbol{\nu}_\mathrm{P} = \boldsymbol{\nu}\,(\lambda, \varphi)\ ,$$

if we put

$$\begin{cases} \Lambda = \lambda + \delta\lambda \\ \Phi = \varphi + \delta\varphi \end{cases} , \tag{5.29}$$

we see that

$$\boldsymbol{\delta} = \boldsymbol{\nu}\,(\Lambda, \Phi) - \boldsymbol{\nu}\,(\lambda, \varphi) = \boldsymbol{\nu}_\lambda \delta\lambda + \boldsymbol{\nu}_\varphi \delta\varphi\ .$$

On the other hand

$$\boldsymbol{\nu}_\lambda = \cos\varphi\, \boldsymbol{e}_\lambda\ , \quad \boldsymbol{\nu}_\varphi = \boldsymbol{e}_\varphi\ ,$$

so that

$$\boldsymbol{\delta} = \cos\varphi\, \delta\lambda\, \boldsymbol{e}_\lambda + \delta\varphi\, \boldsymbol{e}_\varphi$$

and $(\cos\varphi\,\delta\lambda,\ \delta\varphi)$ are respectively the northward and the eastward component of $\boldsymbol{\delta}$, in geodetic literature also denoted as

$$\begin{cases} \eta = \cos\varphi\, \delta\lambda \\ \xi = \delta\varphi \end{cases} . \tag{5.30}$$

Therefore, returning to (5.28) and taking the scalar product with $\boldsymbol{e}_\lambda$ and $\boldsymbol{e}_\varphi$, we get respectively (see (2.85))

$$\begin{cases} \boldsymbol{e}_\lambda \cdot \boldsymbol{\delta} = \eta = -\boldsymbol{e}_\lambda \cdot \nabla T = -\dfrac{1}{(\mathcal{N} + h^*)\cos\varphi}\dfrac{\partial T}{\partial\lambda} \\ \boldsymbol{e}_\varphi \cdot \boldsymbol{\delta} = \xi = \widetilde{\delta} - \boldsymbol{e}_\varphi \cdot \nabla T = \widetilde{\delta} - \dfrac{1}{(\mathcal{M} + h^*)}\dfrac{\partial T}{\partial\varphi} \end{cases} . \tag{5.31}$$

The right hand side of (5.31) is known and therefore such a formula gives (η, ξ), i.e. $(\delta\lambda, \delta\varphi)$ and therefore also (λ, φ) by (5.29). The transformation between (Λ, Φ, W) and (λ, φ, h) is so accomplished, at least in a linear approximation, which amounts to a linear error in coordinates at most of 1 cm, as we have seen in various steps. As a remark, we see that the knowledge of the anomalous potential T (P) is essential to perform our coordinate transformation and it is precisely for this reason that, even studying a geometric topic like heights, we need to know how to compute or at least to use T (P). Finally, we observe that the term $\widetilde{\delta}$ in (5.31) has only recently been introduced (see Betti et al. (2016)) and, though small, it can produce sensible effects in long levelling lines in south-north direction at a relevant height h.

5.4 The Helmert Triad (Λ, Φ, H)

The couple $\Sigma = (\Lambda, \Phi)$ is defined by the relation (5.9), as for the Hotine-Marussi coordinates. The coordinate H, called orthometric height of the point P, is defined as follows: with reference to Fig. 5.1, we take the line of the vertical through P, L_v, and we consider the length of L_v between P and the geoid, i.e. the arc $\widehat{\mathrm{P_0P}}$; then

$$H_{\mathrm{P}} = L_v\left(\widehat{\mathrm{P_0P}}\right) , \tag{5.32}$$

taken positively outside the geoid and with the minus sign inside the geoid. A fast comparison shows that H_{P} is precisely equal to the curvilinear coordinate ℓ_{P} defined in Sect. 5.2.

In spite of its intuitive character, and the fact that for a long time it has been considered as a "natural" coordinate to describe the observation equations of spirit levelling (see Chap. 6), the orthometric height has some subtle properties that have made controversial its use in Geodesy.

The first surprising fact is that the lines of the vertical $\{L_v\}$ *are not* the coordinate lines of H. The family $\{L_v\}$ is used to define H, but its coordinate lines are defined by the condition on the other two coordinates,

$$\Phi = \Phi_0 \text{ (constant)}, \quad \Lambda = \Lambda_0 \text{ constant},$$

namely the lines $\{L_H\}$ coincide again with the family of isozenithals, already seen in Sect. 5.3. A little thought will show that, if we wanted a coordinate system where L_v was a coordinate line, we should have chosen a couple of coordinates (Λ_0, Φ_0) to accompany H or W, that are in fact the astrogeodetic coordinates of the projection of P on the geoid $\mathcal{G}$, along L_v itself.

Now it is obvious that, if we move along L_v, the arc length $d\ell_v$ is

$$d\ell_v = dH \tag{5.33}$$

when we use H, while

$$\mathrm{d}\ell_v = -\frac{\mathrm{d}W}{g} \tag{5.34}$$

when we use W as a parameter. The relations (5.33) and (5.34), especially when they are written in the form

$$\mathrm{d}W = -g\,\mathrm{d}H\ , \tag{5.35}$$

have been already source of controversy because of its imprecise notation. In fact (5.33) and (5.34) are meant to be valid *only* along L_v; they are not equalities between total differentials. Borrowing from an old notation, the relation (5.35), which in general *is wrong*, should be written as

$$\mathrm{d}_n W = -g\,\mathrm{d}H\ , \tag{5.36}$$

meaning that the increment $\mathrm{d}_n W$ is computed exclusively along L_v. In fact, if we move $\boldsymbol{r}$ by $\mathrm{d}\boldsymbol{r} = \boldsymbol{n}\,\mathrm{d}H$ (recall that $\boldsymbol{n} = -\frac{\boldsymbol{g}}{g}$ is always tangent to L_v), we have

$$\mathrm{d}_n W = \boldsymbol{g} \cdot \mathrm{d}\boldsymbol{r} = \boldsymbol{g} \cdot \left(\frac{\boldsymbol{g}}{g}\right) \mathrm{d}H = -g\,\mathrm{d}H\ ,$$

confirming the correctness of (5.36). On the contrary, if $\mathrm{d}\boldsymbol{r}$ is pointing in any direction in space, the relation (5.35) cannot be maintained any more, because, if this would be true, we should have as well

$$\mathrm{d}H = 0 \ \Rightarrow\ \mathrm{d}W = 0\ ,$$

namely equipotential surfaces should have a constant orthometric height too. But in this case

$$g = \frac{\mathrm{d}W}{\mathrm{d}H}$$

should also be constant on an equipotential surface and this is known to be false on an empirical ground; on the other hand, even the normal gravity is not constant on the Earth ellipsoid, which is an equipotential of the normal potential. A deeper analysis (see Sansò and Vanìcek 2006) can show that the only field for which a relation like (5.36) is true is that with a purely spherical potential $\frac{\mu}{r}$.

Now we have to study the transformation of (Λ, Φ, H) into the other geometrical coordinates. As we shall see, to do that we will have to make in any way some hypothesis on the density of topographic masses, e.g. $\rho_{\mathrm{c}} = 2.67\ \mathrm{g\ cm^{-3}}$. This introduces an unavoidable systematic error into the relation between H_{P} and the geodetic observables and therefore into the use of H_{P} itself. In principle we could say that writing (5.5) in the form

$$C(\mathrm{P}) = W_0 - W(\mathrm{P}) = g_\mathrm{P} H_\mathrm{P}\left(1 + \frac{H_\mathrm{P}}{R}\right) - \frac{p}{2} H_\mathrm{P}^2 \tag{5.37}$$

allows to compute $W(\mathrm{P})$ from H_P and therefore, once (Λ, Φ, W) are known, we can repeat the reasoning of Sect. 5.3 to derive (λ, φ, h). This is basically the solution developed by Helmert and we shall shortly report it in a remark at the end of the section. Yet, this is too intricate and, more important, subject to larger errors. We prefer here to go along a way that, exploiting relations between anomalous quantities only, implies smaller errors.

First, inspecting Fig. 5.1, we shall prove what we call the Operative Lemma of Orthometric Heights.

The Operative Lemma of Orthometric Heights: with an accuracy of about 1 cm, or better one can write everywhere on the surface S

$$h_\mathrm{P} = H_\mathrm{P} + N_{\mathrm{P_e}} . \tag{5.38}$$

Proof Looking at Fig. 5.1, (5.38) means

$$\overline{\mathrm{P_eP}} = \widehat{\mathrm{P_0P}} + \overline{\mathrm{P_eP''}} ; \tag{5.39}$$

we prove (5.39) by showing that

$$\widehat{\mathrm{P_0P}} - \overline{\mathrm{P''P}} \cong 0 , \tag{5.40}$$

at the approximation level of 1 cm.

Treating orders of magnitude, we can assume that δ is constant along L_v, because it is known (cf. (3.25)) that the variation of δ along L_v is one order of magnitude smaller than δ itself; moreover we know that $\delta \le 3 \cdot 10^{-4}$.

Then we can write

$$\overline{\mathrm{P'P}} = \int_0^{H_\mathrm{P}} \cos\delta \, \mathrm{d}H \cong H_\mathrm{P}\left(1 - \frac{1}{2}\delta^2\right)$$

$$\overline{\mathrm{P''P'}} = \overline{\mathrm{P_0P'}} \sin\delta \cong H_\mathrm{P} \cdot \delta \cdot \delta = H_\mathrm{P}\,\delta^2 .$$

Therefore

$$\overline{\mathrm{P''P}} \cong H_\mathrm{P} + H_\mathrm{P}\frac{1}{2}\delta^2$$

and then

$$\widehat{\mathrm{P_0P}} - \overline{\mathrm{P''P}} = H_\mathrm{P}\frac{1}{2}\delta^2 ;$$

The above, with $H_{\mathrm{P}} = 6$ km and $\delta = 3 \cdot 10^{-4}$, attains the value of 0.27 mm, which is zero at our approximation level.

Now that (5.38) is proved, we have to show how to compute $N_{\mathrm{P_e}}$ by using anomalous quantities only. We first observe that the fundamental equation of Physical Geodesy, also recalling Bruns' relation (4.82), can be written as

$$-T' + \frac{\gamma'}{\gamma} T = -\gamma \frac{\partial}{\partial h} \frac{T}{\gamma} = -\gamma \frac{\partial}{\partial h} \zeta = \Delta g$$

or

$$\frac{\partial \zeta}{\partial h} = -\frac{\Delta g}{\gamma} . \tag{5.41}$$

Then, integrating (5.41) between P″ and P of Fig. 5.1, we get

$$N - \zeta_{\mathrm{P}} = \int_{\mathrm{P}''}^{\mathrm{P}} \frac{\Delta g}{\gamma} \, \mathrm{d}h \tag{5.42}$$

or, recalling that $\overline{\mathrm{P}''\mathrm{P}} \cong H_{\mathrm{P}}$,

$$N_{\mathrm{P}} = \frac{T(\mathrm{P})}{\gamma} + \int_0^{H_{\mathrm{P}}} \frac{\Delta g}{\gamma} \, \mathrm{d}h . \tag{5.43}$$

If we know $T(\lambda, \varphi, h)$, we can always compute $T(\Lambda, \Phi, H)$ committing an error of 1 cm at most, so the first term in (5.43) is known.

Now we use the identity

$$\int_0^x f(t)\,\mathrm{d}t = x f(x) - \int_0^x t f'(t)\,\mathrm{d}t ,$$

to compute the integral in (5.43). We obtain (remember that $\overline{\mathrm{P}''\mathrm{P}} \cong H_{\mathrm{P}}$)

$$\int_0^{H_{\mathrm{P}}} \frac{\Delta g}{\gamma} \, \mathrm{d}h = H_{\mathrm{P}} \frac{\Delta g_{\mathrm{P}}}{\gamma} - \int_0^{H_{\mathrm{P}}} h \left(\frac{\Delta g}{\gamma} \right)' \mathrm{d}h . \tag{5.44}$$

On the other hand, recalling (5.6) and (5.8),

$$\begin{aligned} \left(\frac{\Delta g}{\gamma} \right)' &= \frac{\gamma \Delta g' - \gamma' \Delta g}{\gamma^2} = \frac{1}{\gamma} \left(-2C_0 \Delta g + q \right) - \frac{\Delta g}{\gamma} \left(-2C_0 - 2\frac{\omega_0^2}{\gamma} \right) = \\ &= \frac{q}{\gamma} + \frac{\Delta g}{\gamma} \frac{2\omega^2}{\gamma} . \end{aligned} \tag{5.45}$$

Now notice that $q \sim 0.2$ mGal m^{-1} and that $\left|\frac{\Delta g}{\gamma}\right| < 10^{-4}$ while $2\omega^2 \cong 10^{-2}$ mGal m^{-1}, so that the second term in (5.45) is five orders of magnitude smaller than the first.

So returning to (5.44), we find

$$\int_0^{H_P} \frac{\Delta g}{\gamma}\, \mathrm{d}h \cong H_P \frac{\Delta g_P}{\gamma} - \int_0^{H_P} q \frac{h}{\gamma}\, \mathrm{d}h \,. \tag{5.46}$$

An easy computation of orders of magnitude shows that (5.46) can amount up to a few meters for $H = 6$ km.

Summarizing we have the solution

$$N_P = \frac{1}{\gamma}\left(T_P + \Delta g_P H_P\right) - \int_0^{H_P} q \frac{h}{\gamma}\, \mathrm{d}h \,, \tag{5.47}$$

which used in (5.38) provides the sought transformation. We note that such a formula, which is now standard in geodetic literature, gives the direct dependence of N_P on the profile of $\rho\,(\mathrm{Q})$ along the vertical of P, through the parameter $q = 4\pi G\rho$.

It is not difficult to see that, by taking $\rho = \mathrm{const}$, a further approximation of (5.47) gives

$$N_P = \frac{1}{\gamma}\left(T_P + \Delta g_P H_P - \frac{1}{2} q\, H_P^2\right) . \tag{5.48}$$

We conclude this section by a remark on the so called Helmert heights.

Remark As commented before, we want to return to the relation between H_P and W_P, which was originally figured out by Helmert, following his definition of orthometric height.

This was derived by the following consideration: start with

$$W_0 - W\,(\mathrm{P}) = -\int_{\mathrm{P}_0}^{\mathrm{P}} \mathrm{d}W = \int_{\mathrm{P}_0}^{\mathrm{P}} g\,\mathrm{d}H \equiv H_P \frac{1}{H_P} \int_0^{H_P} g\,\mathrm{d}H = H_P\,\overline{g} \,, \tag{5.49}$$

where $\overline{g}$ is just the average of g along L_v, between P_0 and P. Then Helmert's reasoning continues with the computation of $\overline{g}$ under the hypothesis that $\rho = \mathrm{const}$ and that g linearly depends on H. But we have already performed this calculation, leading to (5.5), when we disregard the third order term. So we can write

$$C\,(\mathrm{P}) = W_0 - W\,(\mathrm{P}) = g_P\, H_P + \left(\frac{g_P}{R} - \frac{p}{2}\right) H_P^2$$

and then, taking into account that $\frac{H_P}{R} \leq 10^{-3}$, we substitute into the second term of the above formula g_P with the mean value of the normal gravity, $\overline{\gamma}_0$, already computed in Sect. 5.3, Eq. (5.15).

All that gives

$$C\,(\mathrm{P}) = W_0 - W\,(\mathrm{P}) = g_P\, H_P + 0.0424\, H_P^2\;, \tag{5.50}$$

with g in Gal and H_P in km. Eq. (5.50) is exactly what one can find in literature (cf. Heiskanen and Moritz (1967, Eq. 4.4)).

5.5 The Molodensky Triad (λ, φ, h^*)

The coordinates (λ, φ) are taken as the ellipsoidal longitude and latitude and their knowledge implies that of $\boldsymbol{\nu}$, i.e. of the ellipsoidal normal passing through P. The h^* is the normal height, we have already defined in Sect. 4.7, and its defining equation is (see (4.80), (4.81))

$$U\,(\sigma, h^*) \equiv W\,(\sigma, h)\;, \tag{5.51}$$

namely the normal potential at height h^* along the ellipsoidal normal through P should be equal to the actual potential at P. The relation between h and h^* is provided by the Bruns relation (4.82), i.e.

$$h = h^* + \zeta = h^* + \frac{T}{\gamma}\;. \tag{5.52}$$

The relation (5.52) derives, as we know, from a linearization and as such it bears some approximation. In any way, as always, it assumes that we know T in ellipsoidal coordinates. To be precise, the term $\frac{T}{\gamma}$ in (5.52) should be computed with the ellipsoidal height fixed at h^*; so (5.52) gives the transformation from h^* to h. The inverse transformation can always be derived from (5.52), reversed in the form

$$h^* = h - \frac{T}{\gamma}\;, \tag{5.53}$$

where now $\zeta = \frac{T}{\gamma}$ can be computed at the ellipsoidal height $h = h_P$. In fact, since according to (5.41) we have

$$|\zeta\,(h) - \zeta\,(h^*)| \leq \int_{h^*}^{h} \frac{\Delta g}{\gamma}\,\mathrm{d}h \cong \frac{|\Delta g|}{\gamma}\,|\zeta|\;,$$

we see that, with $|\zeta| = 100$ m and $|\Delta g| = 100$ mGal,

$$|\zeta(h) - \zeta(h^*)| \le 1\text{cm} .$$

Would such an accuracy be deemed insufficient, we can always resort to the defining equation (5.51). For that we can use the last equation in Table 3.2, neglecting the last term $\frac{1}{2}\frac{\tau_1^2}{\gamma_0}\frac{h^2}{3}$, which divided by γ_0 is of the order of magnitude of less than 1 μm even for $h = 6$ km; this can be put into the form

$$U_0 - U(h^*) = W_0 - W(\mathrm{P}) = C(\mathrm{P}) = \gamma_e(\varphi)\,h^* - \frac{1}{2}\gamma_1(\varphi)\,h^{*2} + \frac{1}{3}\gamma_2(\varphi)\,h^{*3}, \tag{5.54}$$

with γ in Gal and h^* in km.

Indeed, knowing h and so $W(\mathrm{P})$ and $C(\mathrm{P})$, one can solve (5.54) for h^*; yet, to avoid numerical instabilities, it is convenient to write (5.54) in the form

$$h^* = \frac{C(\mathrm{P})}{\gamma_e(\varphi) - \frac{1}{2}\gamma_1(\varphi)\,h^* + \frac{1}{3}\gamma_2(\varphi)\,h^{*2}} \tag{5.55}$$

and solve it iteratively, starting with $h^* = 0$ at the right hand side.

In case we would like to transform h^* into H or vice versa, one can combine (5.38) and (5.52) to get

$$h^* = H + N - \zeta, \tag{5.56}$$

where $N - \zeta$ can be derived from (5.47) or even (5.48).

References

Betti B., Carrion D., Sacerdote F., Venuti G. (2016). *The observation equation of spirit leveling in Molodensky's context.* In: Sneeuw N., Novàk P., Crespi M., Sansò F. (eds), VIII Hotine-Marussi Symposium on Mathematical Geodesy, International Association of Geodesy Symposia, 142:213–219. Springer, Cham.

Bocchio F. (1981). *An inverse geodetic singularity problem.* Geophysical Journal of the Royal Astronomical Society, 67:181–187.

Heiskanen W.A. and Moritz H. (1967). *Physical geodesy.* Freeman, San Francisco.

Marussi A. (1985). *Intrinsic geodesy.* Springer, Berlin.

Sansò F., Sideris M.G. (2013). *Geoid determination: Theory and methods.* Lecture Notes in Earth System Sciences, Vol. 110. Springer-Verlag, Berlin, Heidelberg.

Sansò F., Vanìcek P. (2006). *The orthometric height and the holonomity problem.* Journal of Geodesy, 80(5):225–232.

Vanícek P. and Krakiwsky E.J. (1986). *Geodesy: The concepts*, 2nd edn. Elsevier, Amsterdam.

Chapter 6
The Relation Between Levelling, Geodetic and Other Unholonomic Heights

6.1 Outline

Spirit levelling is a complex geodetic measurement that combines several elementary steps, already defined in Sect. 4.6 (see also Fig. 4.6), each of them providing a step increment on a short baseline; such increments are then added along a levelling line, joining two stations P and Q. In this way, recalling (4.56), we have a measurement related to the extremes P and Q and to the path connecting them. If we call M a point running along the line, and with the idea to consider each step as infinitesimal, we can write

$$\delta L = \boldsymbol{n}_{\mathrm{M}} \cdot \mathrm{d}\boldsymbol{r}_{\mathrm{M}} \,, \tag{6.1}$$

$$\Delta_{\mathrm{PQ}} L = \int_{\mathrm{P}}^{\mathrm{Q}} \delta L = \int_{\mathrm{P}}^{\mathrm{Q}} \boldsymbol{n} \cdot \mathrm{d}\boldsymbol{r} \,. \tag{6.2}$$

Since the approximate relation

$$\delta L \cong \mathrm{d}H \tag{6.3}$$

holds for a single step, for many practical applications the Eq. (6.2) has been considered as

$$\Delta_{\mathrm{PQ}} L = H_{\mathrm{Q}} - H_{\mathrm{P}} \,, \tag{6.4}$$

especially when levelling is restricted to an area of a few kilometers.

However, since $\mathrm{d}\boldsymbol{r}$ is in general pointing in an almost horizontal direction and certainly not along the vertical, the claim $\delta L = \mathrm{d}H$ is false, as proved and illustrated in Sect. 5.4, so (6.4) is false too; we could say that (6.4) holds with an acceptable approximation only when the path $\overline{\mathrm{PQ}}$ is at most a few kilometers long. So, as we see, there is an intimate connection between levelling observations and geodetic heights, though with some ambiguity that needs to be resolved.

F. Sansò et al., *Geodetic Heights*, Springer Geophysics,
https://doi.org/10.1007/978-3-030-10454-2_6

In Sect. 6.2 we shall first study the relation between $\Delta_{PQ}L$ and dynamic heights. Then in Sect. 6.3 we investigate the normal heights h_P^*, h_Q^*, showing that, contrary to the geodetic tradition, the observation equation in terms of h^* is more natural in that it requires only the knowledge of the anomalous potential T on the surface. On the contrary, in Sect. 6.4, studying the relation between $\Delta_{PQ}L$ and H_P, H_Q, we will show that such observation equation cannot avoid to introduce the knowledge of the topographic masses density; this is in fact intrinsic into the definition of orthometric height. Finally in Sect. 6.5 we shall discuss a different type of unholonomic height, namely the normal orthometric height, that is in fact used by some countries, so that its relation to $\Delta_{PQ}L$ and to other heights needs to be clarified. A final section, Sect. 6.6, of conclusions follows, with recommendations of practical nature.

6.2 The Observation Equation of ΔL in Terms of Dynamic Heights

The definition of dynamic height is (see (5.15))

$$H_P^D = \frac{C\,(P)}{\overline{\gamma}_0}$$

and, as commented in Sect. 5.3, since $\overline{\gamma}_0$ is just a constant, it bears the same information and geometry as the geopotential number $C\,(P) = W_0 - W\,(P)$, as well as the potential $W\,(P)$ itself.

Recalling (6.1) and (6.2), we start our reasoning from

$$\delta L = \boldsymbol{n} \cdot \mathrm{d}\boldsymbol{r} = -\frac{\boldsymbol{g}}{g} \cdot \mathrm{d}\boldsymbol{r} = -\frac{\mathrm{d}W}{g} \tag{6.5}$$

and

$$\Delta_{PQ}L = -\int_P^Q \frac{\mathrm{d}W}{g}\,, \tag{6.6}$$

where the integral is meant to be computed along the levelling line, namely on the Earth surface.

Equation (6.6) can be further elaborated in the following way

$$\Delta_{PQ}L = -\frac{1}{\overline{\gamma}_0}\int_P^Q \frac{\overline{\gamma}_0}{g}\,\mathrm{d}W = -\frac{1}{\overline{\gamma}_0}\int_P^Q \left(\frac{\overline{\gamma}_0 - \gamma + \gamma - g}{g} + 1\right)\mathrm{d}W =$$

$$= \frac{W\,(\mathrm{P}) - W\,(\mathrm{Q})}{\overline{\gamma}_0} + \int_{\mathrm{P}}^{\mathrm{Q}} \frac{\overline{\gamma}_0 - \gamma}{\overline{\gamma}_0}\,\delta L + \int_{\mathrm{P}}^{\mathrm{Q}} \frac{\gamma - g}{\overline{\gamma}_0}\,\delta L\,. \tag{6.7}$$

Now, the difference $g - \gamma$ is called gravity disturbance δg and it is related to T by

$$\delta g = g\,(\mathrm{P}) - \gamma\,(\mathrm{P}) = \boldsymbol{e}_\gamma \cdot \nabla T \cong -\boldsymbol{\nu} \cdot \nabla T = -T'\,; \tag{6.8}$$

it is known that $\mathcal{O}\,(\delta g) \sim \mathcal{O}\,(\Delta g)$, i.e. $\mathcal{O}\left(\frac{\delta g}{\gamma_0}\right) \sim 10^{-4}$. Moreover it is

$$\frac{W\,(\mathrm{P}) - W\,(\mathrm{Q})}{\overline{\gamma}_0} = \frac{C\,(\mathrm{Q}) - C\,(\mathrm{P})}{\overline{\gamma}_0} = H^{\mathrm{D}}_{\mathrm{Q}} - H^{\mathrm{D}}_{\mathrm{P}}\,.$$

So from (6.7) we derive the observation equation

$$\Delta_{\mathrm{PQ}} L = H^{\mathrm{D}}_{\mathrm{Q}} - H^{\mathrm{D}}_{\mathrm{P}} - \int_{\mathrm{P}}^{\mathrm{Q}} \frac{\gamma - \overline{\gamma}_0}{\overline{\gamma}_0}\,\delta L - \int_{\mathrm{P}}^{\mathrm{Q}} \frac{\delta g}{\overline{\gamma}_0}\,\delta L\,. \tag{6.9}$$

A simple evaluation of the orders of magnitude shows that the first integral in the right hand side of (6.9) can amount up to meters per kilometer of height differences, while the second integral is at most one order of magnitude smaller.

6.3 The Observation Equation of ΔL in Terms of Normal Heights

In this case we return to Eq. (6.1) that we rewrite

$$\delta L = \boldsymbol{n} \cdot \mathrm{d}\boldsymbol{r} = -\frac{\boldsymbol{g}}{g} \cdot \mathrm{d}\boldsymbol{r} = -\frac{\mathrm{d}W}{g}\,. \tag{6.10}$$

Now let us go back to (5.28) and write it in the form

$$\boldsymbol{n} = \boldsymbol{\nu} - \frac{1}{\gamma}\,(I - P_\nu)\,\nabla T + \widetilde{\boldsymbol{\delta}}\,, \tag{6.11}$$

where $\widetilde{\boldsymbol{\delta}}$ is given by (5.27) and $I - P_\nu$ is the projection on the horizontal plane, which is orthogonal to $\boldsymbol{\nu}$.

Using (6.11) in (6.10) yields

$$\delta L = \boldsymbol{\nu} \cdot \mathrm{d}\boldsymbol{r} - \frac{1}{\gamma} \nabla T \cdot \mathrm{d}\boldsymbol{r} + \frac{1}{\gamma} \boldsymbol{\nu} \cdot \nabla T \, \boldsymbol{\nu} \cdot \mathrm{d}\boldsymbol{r} + \widetilde{\boldsymbol{\delta}} \cdot \mathrm{d}\boldsymbol{r} =$$
$$= \mathrm{d}h - \frac{\mathrm{d}T}{\gamma} + \frac{T'}{\gamma} \mathrm{d}h + \widetilde{\boldsymbol{\delta}} \cdot \mathrm{d}\boldsymbol{r} \, . \quad (6.12)$$

We will elaborate the term

$$\frac{\mathrm{d}T}{\gamma} = \mathrm{d}\left(\frac{T}{\gamma}\right) - T \, \mathrm{d}\left(\frac{1}{\gamma}\right) = \mathrm{d}\zeta + T \, \frac{\nabla\gamma \cdot \mathrm{d}\boldsymbol{r}}{\gamma^2} \, . \quad (6.13)$$

To continue our reasoning on (6.13) we use a simplified version of $\frac{\nabla\gamma}{\gamma} \cong \frac{\nabla\gamma}{\gamma_0}$, namely (see Table 3.2 in Sect. 3.5)

$$\frac{\nabla\gamma}{\gamma_0} \cong -3 \cdot 10^{-4} \, \boldsymbol{\nu} + 0.8 \cdot 10^{-6} \sin 2\varphi \, \boldsymbol{e}_\varphi \, ,$$

and observe that

$$\mathcal{O}\left(\frac{T}{\gamma} \frac{\boldsymbol{\nu} \cdot \nabla\gamma}{\gamma}\right) \cong \mathcal{O}\left(\frac{T}{\gamma} \frac{\gamma'}{\gamma}\right) \cong 3 \cdot 10^{-5} \, ,$$

$$\mathcal{O}\left(\frac{T}{\gamma} \frac{|(I - P_\nu) \, \nabla\gamma|}{\gamma}\right) \cong \mathcal{O}\left(\frac{T}{\gamma} \frac{\left|\frac{1}{R} \frac{\partial\gamma}{\partial\varphi}\right|}{\gamma}\right) \sim 10^{-7} \, .$$

So the effect of the horizontal component of the term $\frac{T}{\gamma} \frac{\nabla\gamma}{\gamma} \cdot \mathrm{d}\boldsymbol{r}$, integrated over a 100 km line, is at most 1 cm, while the effect of the vertical component is 3 ΔL cm (ΔL in km), i.e. with a rise of 6 km along the line it can go up to 18 cm.

Therefore the vertical component of this term should be accounted for, especially in mountainous areas, while the horizontal one can be neglected. Therefore, returning to (6.13), we get

$$\frac{\mathrm{d}T}{\gamma} \cong \mathrm{d}\zeta + \frac{T}{\gamma} \frac{\gamma'}{\gamma} \mathrm{d}h \, ,$$

which, in (6.12), yields

$$\delta L = \mathrm{d}h - \mathrm{d}\zeta - \frac{T}{\gamma} \frac{\gamma'}{\gamma} \mathrm{d}h + \frac{T'}{\gamma} \mathrm{d}h + \widetilde{\boldsymbol{\delta}} \cdot \mathrm{d}\boldsymbol{r} =$$
$$= \mathrm{d}\,(h - \zeta) - \frac{1}{\gamma}\left(-T' + \frac{\gamma'}{\gamma} T\right) \mathrm{d}h + \widetilde{\boldsymbol{\delta}} \cdot \mathrm{d}\boldsymbol{r} =$$
$$= \mathrm{d}h^* - \frac{\Delta g}{\gamma} \mathrm{d}h + \widetilde{\boldsymbol{\delta}} \cdot \mathrm{d}\boldsymbol{r} \, . \quad (6.14)$$

This is the sought observation equation of the levelling increment over one step; then $\Delta_{PQ}L$ has the observation equation

$$\Delta_{PQ}L = h_Q^* - h_P^* - \int_P^Q \frac{\Delta g}{\gamma}\,\mathrm{d}h + \int_P^Q \widetilde{\boldsymbol{\delta}} \cdot \mathrm{d}\boldsymbol{r} \,. \tag{6.15}$$

As promised, all the terms in (6.15) can be computed from surface anomalous quantities. In particular the term

$$\mathrm{DC} = \int_P^Q \frac{\Delta g}{\gamma}\,\mathrm{d}h \,,$$

also known in literature as dynamic correction, can amount up to 10^{-4} times the levelling increment, namely to several dozens of cm if P is by the sea and Q is on a high mountain.

On the contrary the last term, only recently reported in literature (see Betti et al. 2016),

$$\int_P^Q \widetilde{\boldsymbol{\delta}} \cdot \mathrm{d}\boldsymbol{r} \cong 5.3 \cdot 10^{-3} \int_P^Q \sin 2\varphi \,\frac{h}{a}\,(\mathcal{M} + h)\,\mathrm{d}\varphi \,,$$

can obviously give a sensible contribution only for a levelling line at altitude and developing in the north-south direction. For instance a levelling line on the Andes, 60 km long, around $\varphi = -45°$, at an altitude of 2 km, will have a correction term $\int_P^Q \widetilde{\boldsymbol{\delta}} \cdot \mathrm{d}\boldsymbol{r}$ of about 10 cm.

All in all, we have shown that by calling NC, normal correction, the term

$$\mathrm{NC} = \int_P^Q \frac{\Delta g}{\gamma}\,\mathrm{d}h - \int_P^Q \widetilde{\boldsymbol{\delta}} \cdot \mathrm{d}\boldsymbol{r} \,, \tag{6.16}$$

the levelling increment has observation equation

$$\Delta_{PQ}L = h_Q^* - h_P^* - \mathrm{NC} \,, \tag{6.17}$$

where the last term can be effectively computed by surface quantities. This means, for instance, that nowadays NC can be computed to a sufficient degree of accuracy from some global model of T, e.g. from EGM2008; note that, on the contrary, such a model could not be used to compute quantities inside the masses, where one should use the methods explained in Sect. 5.2.

Remark One possible objection to the computability of NC is that in principle the expression (6.16) should be reckoned along the "true" levelling line, the profile of which should therefore be known to compute the DC, while the second term is affected in any circumstance by a negligible error. Yet we can respond that assuming that the levelling line is known with some 10 m errors in height, what nowadays is easy to achieve e.g. by Real Time Kinematic GNSS observations, then Δg is known with at most 1 mGal error, implying that DC can be computed with an error of $10^{-6}\ \Delta_{\mathrm{PQ}}L$. This is certainly negligible; a similar consideration holds for the term $\int_{\mathrm{P}}^{\mathrm{Q}} \widetilde{\delta} \cdot \mathrm{d}\boldsymbol{r}$.

The conclusion of this section is that levelling networks should be compensated, after the application of normal corrections, directly in terms of normal heights and, to this aim, the use of global models to compute normal corrections can give accurate enough results, with particular caution in areas of rough topography.

6.4 The Observation Equation of ΔL in Terms of Orthometric Heights

The wanted observation equation is easily derived from (6.16) and (6.17), taking into account the following elementary relation, which takes advantage of (5.38) and (5.43),

$$\begin{aligned} h_{\mathrm{Q}}^* - h_{\mathrm{P}}^* &= h_{\mathrm{Q}} - h_{\mathrm{P}} - \zeta_{\mathrm{Q}} + \zeta_{\mathrm{P}} = \\ &= \left(h_{\mathrm{Q}} - N_{\mathrm{Q}}\right) - \left(h_{\mathrm{P}} - N_{\mathrm{P}}\right) + \left(N_{\mathrm{Q}} - \zeta_{\mathrm{Q}}\right) - \left(N_{\mathrm{P}} - \zeta_{\mathrm{P}}\right) = \\ &= H_{\mathrm{Q}} - H_{\mathrm{P}} + \int_{\mathrm{Q}_0}^{\mathrm{Q}} \frac{\Delta g}{\gamma}\, \mathrm{d}h - \int_{\mathrm{P}_0}^{\mathrm{P}} \frac{\Delta g}{\gamma}\, \mathrm{d}h\ , \end{aligned} \tag{6.18}$$

where Q_0 and P_0 are the projections of Q and P, respectively, on the geoid.

Defining the orthometric correction OC as

$$\mathrm{OC} = \mathrm{NC} - \int_{\mathrm{Q}_0}^{\mathrm{Q}} \frac{\Delta g}{\gamma}\, \mathrm{d}h + \int_{\mathrm{P}_0}^{\mathrm{P}} \frac{\Delta g}{\gamma}\, \mathrm{d}h \tag{6.19}$$

and substituting (6.18) and (6.19) into (6.17), we get

$$\Delta_{\mathrm{PQ}}L = H_{\mathrm{Q}} - H_{\mathrm{P}} - \mathrm{OC}\ . \tag{6.20}$$

As we see, contrary to the case of the normal correction, in (6.19) we find that the orthometric correction cannot be computed without making hypotheses on the density of topographic masses. In fact, recalling (5.46), we can also write

$$\mathrm{OC} = \int_{\mathrm{P}}^{\mathrm{Q}} \frac{\Delta g}{\gamma}\, \mathrm{d}h - H_{\mathrm{Q}} \frac{\Delta g_{\mathrm{Q}}}{\gamma} + H_{\mathrm{P}} \frac{\Delta g_{\mathrm{P}}}{\gamma} + \int_0^{H_{\mathrm{Q}}} q \frac{h}{\gamma}\, \mathrm{d}h - \int_0^{H_{\mathrm{P}}} q \frac{h}{\gamma}\, \mathrm{d}h \,, \tag{6.21}$$

where $q = 4\pi G\rho$. Note that in (6.21) the first integral is along the levelling line, while the last two are inside the masses.

Since here we are reasoning apart from measurement errors, we could say that (6.21) can establish an orthometric coordinate system in a certain area if starting from a point P_0 on the geoid, we could reach every point Q in the area, connecting it to P_0 by a levelling line. In this case, noting that $\mathrm{P} = \mathrm{P}_0$, $H_{\mathrm{P}_0} = 0$, we have from (6.16) and (6.17)

$$H_{\mathrm{Q}} = \Delta_{\mathrm{P}_0\mathrm{Q}} L + \int_{\mathrm{P}_0}^{\mathrm{Q}} \frac{\Delta g}{\gamma}\, \mathrm{d}h - H_{\mathrm{Q}} \frac{\Delta g_{\mathrm{Q}}}{\gamma} + \int_0^{H_{\mathrm{Q}}} q \frac{h}{\gamma}\, \mathrm{d}h \,; \tag{6.22}$$

this explicit formula is fundamental to understand the next remark.

Remark In geodetic literature it is often written that the relation (5.38)

$$h = H + N$$

can be used to assess the accuracy of a gravimetric geoid, by comparing it with the difference $h - H$, where h can be obtained by GNSS measurements, while H can be obtained by levelling.

We claim that the statement is wrong, at least at the level of accuracy of one centimeter. In fact we know from Sect. 4.7 that surface gravimetric data can provide only T, from the telluroid upward, and from this the height anomaly can be computed via Bruns' relation, $\zeta = \dfrac{T}{\gamma}$; so N can be derived only by making hypotheses on the topographic masses, e.g. by the (approximate) relation (5.47)

$$N_{\mathrm{Q}} = \frac{1}{\gamma}\left(T_{\mathrm{Q}} + \Delta g_{\mathrm{Q}} H_{\mathrm{Q}}\right) - \int_0^{H_{\mathrm{Q}}} q \frac{h}{\gamma}\, \mathrm{d}h \,.$$

On the other hand the above relation clearly shows that an error $\varepsilon(q)$, due to an error in ρ, would cause in the computation of N an error $\varepsilon(N)$ given by

$$\varepsilon(N) = -\int_0^{H_{\mathrm{Q}}} \varepsilon(q) \frac{h}{\gamma}\, \mathrm{d}h \,,$$

while the same error in q would generate in H an error exactly equal in modulus but opposite in sign, so that the relation (5.38) can continue to hold, despite the fact that both H and N are affected by errors. As for the order of magnitude of such an error, one can use the rough appraisal

$$|\varepsilon(N)| = -\mathcal{O}\left(\int_0^{H_Q} \varepsilon(q)\frac{h}{\gamma}\,\mathrm{d}h\right) \sim \frac{2\pi G\,\varepsilon(\rho)\,H^2}{\gamma_0};$$

therefore, with an error of 10% in ρ, this would give

$$|\varepsilon(N)| \sim 10^{-5}H^2\,, \quad (H \text{ in km})\,,$$

which is 1 cm at $H = 1$ km, but 4 cm at $H = 2$ km, and so forth. So we expect that, in particular in mountainous areas, both H and N might be affected by centimetric errors without that (5.38) could reveal it.[1]

We think that the right approach would be to evaluate normal heights directly from levelling, as explained in Sect. 6.3, and then the height anomalies derived by some solution T of the GBVP, to be tested with the relation

$$h = h^* + \zeta\,,$$

where all terms can be observed and computed independently.

6.5 Levelling and Normal Orthometric Heights: An Unholonomic Coordinate

In a sense an unholonomic coordinate is a contradiction in terms, in that it is not a function of a point, as we defined it in Sect. 2.2, but rather a function of a point and a path, as it happens when we make line integrals of non-exact differential forms.

We shall deviate here from the approach of the previous sections and, instead of starting from the observation equation of δL or ΔL, we shall rather start from the other side, namely the definition of normal orthometric heights.

Borrowing for instance from the "Geodetic Glossary" of the National Geodetic Survey, we define the normal orthometric height, H^{no}, as

$$H_{\mathrm{Q}}^{\mathrm{no}} = \frac{1}{\overline{\gamma}^{\mathrm{no}}}\int_{\mathrm{P}_0}^{\mathrm{Q}} \gamma\,\delta L\,, \qquad (6.23)$$

[1]The authors are aware that while editing the book the same result has been independently published by (Sjoberg 2018); we are then happy to acknowledge this coincidence, confirming our findings.

where P_0 is an emanation point on the geoid and $\overline{\gamma}^{no}$ is the mean value of γ along the ellipsoidal normal, up to H^{no} itself, i.e.

$$\overline{\gamma}^{no} = \frac{1}{H_Q^{no}} \int_0^{H_Q^{no}} \gamma(z)\, dz \, . \tag{6.24}$$

Indeed $\overline{\gamma}^{no}$ is a function of point and in fact, using the approximate formulas of Table 3.2, we can even give its explicit form, namely

$$\overline{\gamma}^{no} = \gamma_e(\varphi) - \frac{1}{2}\gamma_1(\varphi)\, H^{no} + \frac{1}{3}\gamma_2(\varphi)\,(H^{no})^2 \, . \tag{6.25}$$

Indeed, as it already happened with Helmert's definition of orthometric height, (6.24) is an implicit equation for H^{no}.

On the other hand H^{no} is not a holonomic coordinate because

$$\gamma\, \delta L = -\frac{\gamma}{g}\, dW \tag{6.26}$$

is certainly not an exact differential: in fact $\frac{\gamma}{g}$ is not constant on equipotential surfaces. As a matter of fact, even going from P_0 to another point Q_0 on the geoid, we are not sure to find $H_{Q_0}^{no} = 0$.

Yet the rationale behind (6.23) as a substitute of H_Q, is that, as nicely stated by D. Hock (private communication), "at least the average variations of gravity due to latitude and height effect was considered, while the irregular variations of the gravity field had been neglected".

The integral in (6.23) is called spheropotential number, C', and for it one has

$$C' = \int_{P_0}^{Q} \gamma\, \delta L = \int_{P_0}^{Q} (\gamma - g)\, \delta L + \int_{P_0}^{Q} g\, \delta L \, . \tag{6.27}$$

On the other hand, as we have already seen in Sect. 6.2, it is

$$g - \gamma = \boldsymbol{e}_\gamma \cdot \nabla T \cong -\boldsymbol{\nu} \cdot \nabla T = -T' = \delta g \, .$$

Moreover

$$g\, \delta L = -\boldsymbol{g} \cdot d\boldsymbol{r} = -dW \, ,$$

so that (6.27) becomes

$$C' = C + \int_{P_0}^{Q} \delta g\, \delta L$$

and (6.23) reads

$$H^{\text{no}} = \frac{C}{\overline{\gamma}^{\text{no}}} - \frac{1}{\overline{\gamma}^{\text{no}}} \int_{\mathrm{P}_0}^{\mathrm{Q}} \delta g \, \delta L \, . \tag{6.28}$$

Now consider that, by definition of normal height,

$$C = W_0 - W(h) = U_0 - U\left(h^*\right) = -\int_0^{h^*} \gamma \cdot \boldsymbol{\nu} \, \mathrm{d}h \cong$$

$$\cong \int_0^{h^*} \gamma \, \mathrm{d}h = h^* \frac{1}{h^*} \int_0^{h^*} \gamma \, \mathrm{d}h \equiv h^* \overline{\gamma}^* \, ; \tag{6.29}$$

here we have denoted by $\overline{\gamma}^*$ the mean of γ between 0 and h^*.

We anticipate that $\delta H^{\text{no}} = H^{\text{no}} - h^*$ is certainly smaller than 1 m, therefore we see from (6.25), keeping only the main term in γ_1 which is enough for the present calculation, that

$$\left|\overline{\gamma} - \overline{\gamma}^*\right| \lessapprox 0.15 \text{ Gal km}^{-1} \cdot 10^{-3} \text{ km} = 1.5 \cdot 10^{-3} \text{ Gal} \, .$$

Therefore

$$\frac{C}{\overline{\gamma}^{\text{no}}} = \frac{\overline{\gamma}^*}{\overline{\gamma}^{\text{no}}} h^* = h^* + \frac{\overline{\gamma}^* - \overline{\gamma}^{\text{no}}}{\overline{\gamma}^{\text{no}}} h^* \, ,$$

where the last term is of the order of $1.5 \; 10^{-7} \; h^*$, i.e. less than 1 mm even for $h^* = 6\,\text{km}$. So we can put

$$\frac{C}{\overline{\gamma}^{\text{no}}} \sim h^* \, ,$$

to find from (6.28)

$$H^{\text{no}} = h^*_{\mathrm{Q}} - \frac{1}{\overline{\gamma}^{\text{no}}} \int_{\mathrm{P}_0}^{\mathrm{Q}} \delta g \, \delta L \, . \tag{6.30}$$

With (6.30) we can verify a posteriori that our guess that $H^{\text{no}} - h^*$ is less than 1 m is correct; in fact $\mathcal{O}\left(\int_{\mathrm{P}_0}^{\mathrm{Q}} \delta g \, \delta L\right) \sim 10^{-4} \, \Delta L$, i.e. 60 cm for $\Delta L = 6\,\text{km}$!

On a theoretical ground, (6.30) shows that there cannot be much advantage in using H^{no} instead of h^*. Yet, for the sake of completeness, let us further develop (6.30) to find the relation between H^{no} and the levelling observable $\Delta_{\mathrm{P}_0\mathrm{Q}} L$.

Going back to (6.17) and observing that when $\mathrm{P}_0 \in \mathcal{G}$ then $h^*_{\mathrm{P}_0} = 0$, we see that

$$h_{\mathrm{Q}}^{*} = \Delta_{\mathrm{P_0Q}}L + \int_{\mathrm{P_0}}^{\mathrm{Q}} \frac{\Delta g}{\gamma}\,\mathrm{d}h - \int_{\mathrm{P_0}}^{\mathrm{Q}} \widetilde{\boldsymbol{\delta}} \cdot \mathrm{d}\boldsymbol{r}\,.$$

Using this relation in (6.30), we get

$$H^{\mathrm{no}} = \Delta_{\mathrm{P_0Q}}L + \int_{\mathrm{P_0}}^{\mathrm{Q}} \frac{\Delta g}{\gamma}\,\mathrm{d}h - \int_{\mathrm{P_0}}^{\mathrm{Q}} \frac{\delta g}{\overline{\gamma}^{\mathrm{no}}}\,\delta L - \int_{\mathrm{P_0}}^{\mathrm{Q}} \widetilde{\boldsymbol{\delta}} \cdot \mathrm{d}\boldsymbol{r}\,. \tag{6.31}$$

The integral of $\frac{\delta g}{\overline{\gamma}^{\mathrm{no}}}\,\delta L$ can indeed be transformed into an integral of $\frac{\delta g}{\overline{\gamma}^{\mathrm{no}}}\,\mathrm{d}h$ because $\frac{\delta g}{\overline{\gamma}^{\mathrm{no}}}$ is already of a maximum order of 10^{-4}. Moreover by writing

$$\frac{\Delta g}{\gamma} - \frac{\delta g}{\overline{\gamma}^{\mathrm{no}}} = \frac{\Delta g - \delta g}{\gamma} + \left(\frac{1}{\gamma} - \frac{1}{\overline{\gamma}^{\mathrm{no}}}\right)\delta g \cong \frac{\gamma'}{\gamma}\frac{T}{\gamma} - \frac{\gamma - \overline{\gamma}^{\mathrm{no}}}{\gamma}\frac{\delta g}{\gamma}\,,$$

we easily verify that the integral in dh of the last term is irrelevant, so that (6.31) becomes

$$H^{\mathrm{no}} = \Delta_{\mathrm{P_0Q}}L + \int_{\mathrm{P_0}}^{\mathrm{Q}} \frac{\gamma'}{\gamma}\,\zeta\,\mathrm{d}h - \int_{\mathrm{P_0}}^{\mathrm{Q}} \widetilde{\boldsymbol{\delta}} \cdot \mathrm{d}\boldsymbol{r}\,. \tag{6.32}$$

A fast evaluation of the orders of magnitude of the correction terms in (6.32) shows that in general these are smaller than NC or OC, yet the price to pay in using (6.32) is that the so calculated value does depend on the path between $\mathrm{P_0}$ and Q because H^{no} is unholonomic. A recent study with a precise numerical evaluation of the effects of using H^{no} in Australia, i.e. a nation that has officially adopted a normal orthometric height system (Featherstone and Kuhn 2006), can be found in Filmer et al. (2010).

6.6 Conclusions

Since the matter has a relevant practical impact on the adoption of national height systems for geodetic purposes, we like to summarize the relevant conclusions that one can draw from the discussions of the chapters:

1. levelling measurements should always be accompanied by corrections that depend on the height coordinate chosen,
2. to compute corrections we need an approximate position of levelling stations, say with 10 m accuracy; this is easily achievable by RTK observations that should always accompany levelling, especially in mountainous areas,

3. all corrections involving the knowledge of the anomalous potential on the Earth surface and outside can be computed to a sufficient accuracy by a good global model of T; obviously a good local model of T will do a better job,
4. the use of orthometric heights implies the application of the OC which depends on the knowledge of density of topographic masses; since such a detailed knowledge is usually not available, the OC can be computed only by making hypotheses on ρ, like $\rho = 2.67$ g cm^{-3}, which can imply a systematic error up to several centimeters, especially in mountainous areas,
5. normal heights, with their effectively computable NC, seem to be the most natural coordinates to compensate levelling networks; moreover they are consistent with the theory of the GBVP, which is one root of the foundations of Physical Geodesy,
6. the relation

$$h = H + N$$

should not be used to assess the accuracy of the estimated geoid at centimetric level, because H and N can hide errors equal in modulus and opposite in sign up to several centimeters; rather the relation

$$h = h^* + \zeta$$

can be used to assess the accuracy of the quasi-geoid, ζ, with systematic errors below the centimeter.

References

Betti B., Carrion D., Sacerdote F., Venuti G. (2016). *The observation equation of spirit leveling in Molodensky's context*. In: Sneeuw N., Novàk P., Crespi M., Sansò F. (eds), VIII Hotine-Marussi Symposium on Mathematical Geodesy, International Association of Geodesy Symposia, 142:213–219. Springer, Cham.

Featherstone W.E., Kuhn M. (2006). *Height systems and vertical datums: a review in the Australian context*. Journal of Spatial Science, 51(1):21–42.

Filmer M.S., Featherstone W.E., Kuhn M. (2010). *The effect of EGM2008-based normal, normal-orthometric and Helmert orthometric height systems on the Australian levelling network*. Journal of Geodesy, 84(8):501–513.

Sjoberg L. (2018). *On the topographic bias and density distribution in modelling the geoid and orthometric heights*. Journal of Geodetic Science 8:30–33.

Chapter 7
The Height Datum Problem

7.1 Outline

Normal and orthometric heights are among the most widespread height coordinate systems in use for geodetic purposes. Yet in principle they can be determined only by ground gravimetric measurements combined with levelling so that $W(\mathrm{P})$ becomes available. Nevertheless, what the above measurements can really provide are at most potential differences, $W(\mathrm{P}_0) - W(\mathrm{P})$, for instance with respect to an origin point P_0 of which however the absolute value $W(\mathrm{P}_0)$ is unknown. When P_0 is a tide gauge, we know that we can assume $W(\mathrm{P}_0) \sim W_0$ with an error δW_0 such that $\left|\frac{\delta W_0}{\gamma}\right| < 2\,\mathrm{m}$ (cfr. Sect. 4.6); when P_0 is a point of known ellipsoidal height, e.g. a GNSS permanent station, we can always assume that $h^* \cong \widetilde{h}^* = h - \frac{T_b}{\gamma}$, where T_b is some global model that has been computed with biases and so it has an error which however is almost surely included in the above range.

In oceanic areas the information from radar altimetry and oceanography can be transformed into potential and gravity, yet biases seem to be pervasive and we can only say, after linearization and inversion, that we know $\Delta g + \frac{\gamma'}{\gamma}\delta W$, with the bias δW unknown for large portions of ocean where altimetric tracks can be readjusted at the crossovers (see Sansò and Sideris 2013, Chap. 9).

All in all we can say that instead of knowing $W(\mathrm{P})$, with known horizontal coordinates of P, $\sigma_{\mathrm{P}} = (\lambda_{\mathrm{P}}, \varphi_{\mathrm{P}})$, we rather have the information

$$\widetilde{C}_k(\mathrm{P}) = W(\mathrm{P}_{0k}) - W(\mathrm{P}) = W_{0k} - W(\mathrm{P}) \ , \tag{7.1}$$

which is valid for an area A_k where levelling on land, or track adjustment on ocean, are well connected to some origin P_{0k}.

Assuming for the sake of simplicity that P_{0k} is in any way close to the sea surface, we could say that in A_k we have the approximate potential

F. Sansò et al., *Geodetic Heights*, Springer Geophysics,
https://doi.org/10.1007/978-3-030-10454-2_7

$$\widetilde{W}(\text{P}) = W_0 - \widetilde{C}_k(\text{P}) = W_0 - W(\text{P}_{0k}) + W(\text{P}) \equiv \delta W_{0k} + W(\text{P}) \quad \text{P} \in A_k\,; \quad (7.2)$$

so δW_{0k} has the meaning of the bias of the known $\widetilde{W}(\text{P})$ in the area A_k. Putting together all the areas A_k, that we assume to cover the whole Earth sphere, we can represent our data as an approximate potential

$$\widetilde{W}(\text{P}) = W(\text{P}) + \delta W(\text{P}) \ , \quad (7.3)$$

where

$$\delta W(\text{P}) = \sum_{k=1}^{K} \delta W_{0k}\, \chi_k(\text{P}) \quad (7.4)$$

and

$$\chi_k(\text{P}) = \begin{cases} 1 & \text{P} \in A_k \\ 0 & \text{P} \notin A_k \end{cases} . \quad (7.5)$$

At this point we do not have anymore the telluroid S^*, i.e. we are not able to compute h_P^* by solving (4.80), but we can only put

$$\widetilde{W}(\text{P}) = W(\sigma, h_\sigma) + \delta W(\sigma) = U\left(\sigma, \widetilde{h}_\sigma^*\right) \ , \quad (7.6)$$

so deriving an approximate, or biased, telluroid $\widetilde{S} = \left\{h = \widetilde{h}_\sigma^*\right\}$, such that

$$\text{D}W = W - \widetilde{W} = -\delta W(\sigma) \neq 0\,. \quad (7.7)$$

Accordingly, following the same linearization process as in Sect. 4.7 and recalling (4.78), we arrive at a BVP for the unknown anomalous potential T of the form

$$\begin{cases} \Delta T = 0 & \text{in } \widetilde{\Omega} \\ -T' + \dfrac{\gamma'}{\gamma} T \Big|_{\widetilde{S}} = \text{D}g - \dfrac{\gamma'}{\gamma}\delta W & \text{on } \widetilde{S} \\ T = \mathcal{O}\left(\frac{1}{r^3}\right) \end{cases} . \quad (7.8)$$

Notice that $\text{D}g = g(\text{P}) - \gamma\left(\widetilde{h}^*\right)$ is as a matter of fact what we can compute from gravimetry and the known approximate telluroid $\widetilde{S}$.

As we can see, (7.8) contains the K unknown parameters $\{\delta W_{0k}\}$, so that we can arrive to determine T and $\{\delta W_{0k}\}$ only by means of additional information; we will see in the chapter that this can be provided by points P where both $\widetilde{h}_\text{P}^*$ and h_P are known, to be precise at least one point per patch A_k, although knowing more can indeed improve the accuracy of the solution.

Let us note that, once $\{\delta W_{0k}\}$ are known, the potential $W(\text{P})$ can be retrieved by

$$W(\text{P}) = \widetilde{W}(\text{P}) - \delta W(\text{P})$$

and, since T is also known now without biases, we can return to compute all the transformations already studied in Chap. 5.

The solution of (7.8) is called the unification of the height datum problem, or more precisely, of the global height datum problem. In fact, if we consider as "height datum" the equipotential surface used as origin of orthometric heights, namely the geoid, we see that $\frac{\delta W(\mathrm{P})}{\gamma}$ can be interpreted as the separation between $\widetilde{S}$, which is composed by pieces of equipotential surfaces passing through P_{0k}, and the geoid, where $W(\mathrm{P})$ attains the value W_0. So knowing δW_{0k} means also to be able to transform local orthometric heights, referred to the equipotential through P_{0k}, into true orthometric heights, referred to the geoid.

An important point in the application of the above theory is that, when many points of known ellipsoidal height are present in the same patch A_k, one is led to use a least-squares adjustment to best estimate the $\{\delta W_{0k}\}$. However this requires that the covariance structure of the observations is known. This is particularly complicated for the oceanic areas where data have undergone a deep transformation process. On the other hand, we have already observed at the end of Sect. 4.7 that local models of T are available on continental areas with an overall error r.m.s. at centimetric level in geoid, in the area A of interest. This introduces the possibility of adjusting δW_{0k} for limited areas only, particularly continental areas, avoiding the problem of assigning a stochastic structure to the data in the ocean.

The whole subject of the unification of the height datum is still object of research and not completely assessed. So, in this chapter we aim at presenting the theory and evaluating the error budget with the purpose of demonstrating its feasibility. Some numerical examples, simulated or realistic, are also presented.

7.2 Formulation of the Global Unification of the Height Datum

As explained in the previous section, this problem is a combination of the solution of a GBVP with unknown additional parameters, $\{\delta W_{0k}\}$, and a set of additional data, corresponding to points P_i (at least one per patch A_k) where the ellipsoidal height $h_i = h(\mathrm{P}_i)$ has been observed.

As for the GBVP part, this has already been discussed in Sect. 7.1, leading to the formulation (7.8). Here we underline only that we know from the discussion of Sect. 4.7 that a linear solving operator exists, such that (7.8) can be written as

$$T = \widetilde{\mathcal{S}}\left(\mathrm{D}g - \frac{\gamma'}{\gamma}\delta W\right) = \widetilde{\mathcal{S}}(\mathrm{D}g) - \widetilde{\mathcal{S}}\left(\frac{\gamma'}{\gamma}\delta W\right) ; \tag{7.9}$$

note that here the tilde stems from the fact that we solve with respect to the approximate surface $\widetilde{S}$. Let us observe that the operator $\widetilde{\mathcal{S}}$ is well defined when acting on functions in L^2_σ; this is the case in (7.9), also for the second term in the right hand side, because δW as a piecewise constant function is certainly in L^2_σ.

In this section we develop the theory as if the global model $\widetilde{S}\,(\mathrm{D}g)$ would be really available, given that Dg is the only "observable" quantity at ground level available to us. Actually this is not the case with existing global models, in particular EGM2008. In fact space geodetic techniques, especially in the last two decades with the satellite gravimetry/gradiometry missions CHAMP, GRACE and GOCE, have provided an independent and direct information on the low degrees of the harmonic coefficients of $T\,(\mathrm{P})$; however this issue will be treated separately in the next section.

So we assume to know a biased anomalous potential

$$T_b\,(\mathrm{P}) = \widetilde{S}\,(\mathrm{D}g)\ . \tag{7.10}$$

Subsequently, introducing (7.4) into (7.9), we arrive at the equation

$$T\,(\mathrm{P}) = T_b\,(\mathrm{P}) - \sum_{k=1}^{K} \delta W_{0k}\, \widetilde{S}\left(\frac{\gamma'}{\gamma}\,\chi_k\right)\ , \tag{7.11}$$

with $T_b\,(\mathrm{P})$ known by hypothesis; for later use we can put $F_k\,(\mathrm{P}) = \widetilde{S}\left(\frac{\gamma'}{\gamma}\,\chi_k\right)$, so that (7.11) is rewritten as

$$T\,(\mathrm{P}) = T_b\,(\mathrm{P}) - \sum_{k=1}^{K} \delta W_{0k}\, F_k\,(\mathrm{P})\ . \tag{7.12}$$

Let us consider now the observed $\{h\,(\mathrm{P}_i)\}$, $\mathrm{P}_i \in A_k$; recalling (7.2), we can write

$$\widetilde{W}\,(\mathrm{P}_i) = \delta W_{0k} + W\,(\mathrm{P}_i) = \delta W_{0k} + U\,(h_i) + T\,(\mathrm{P}_i) \qquad \mathrm{P}_i \in A_k\ . \tag{7.13}$$

On the other hand we have, according to (7.6),

$$\widetilde{W}\,(\mathrm{P}_i) = U\left(\widetilde{h}_i^*\right) \tag{7.14}$$

and indeed $\widetilde{h}_i^* = \widetilde{h}^*\,(\mathrm{P}_i)$ is known by hypothesis too. The practical situation is that, if P_i is a geodetic space station, this is connected to the local levelling line, so that $\widetilde{h}_i^*$ is directly known. Putting (7.13) and (7.14) together gives

$$U\,(h_i) - U\left(\widetilde{h}_i^*\right) = -T\,(\mathrm{P}_i) - \delta W_{0k}\ ,$$

which, linearized with respect to $h_i - \widetilde{h}_i^*$, yields

$$h_i - \widetilde{h}_i^* = \frac{T}{\gamma} + \frac{\delta W_{0k}}{\gamma}\ . \tag{7.15}$$

Notice that, as customary, in (7.15) we do not write explicitly where to compute T and γ, because choosing either h_i or $\widetilde{h}_i^*$, in the right hand side of this relation, produces only second order variations. Finally, introducing (7.12) into (7.15), we find

$$h_i - \widetilde{h}_i^* = \frac{T_b(\mathrm{P}_i)}{\gamma} - \frac{1}{\gamma}\sum_{j=1}^{K} \delta W_{0j}\, F_j(\mathrm{P}_i) + \frac{\delta W_{0k}}{\gamma} \qquad \mathrm{P}_i \in A_k\,. \tag{7.16}$$

As we see if we complement (7.16) with the proper error models for h_i, $\widetilde{h}_i^*$ and $T_b(\mathrm{P}_i)$, we have reduced the solution of our unification problem to that of a least-squares system.

Note should be taken that the functions $F_j(\mathrm{P})$ are generally small outside A_j, so (7.16) could become badly conditioned if one of the patches would be void of points P_i where h_i is known, as already stated before. It has to be stressed too that indeed the system (7.16) should be solved for all P_i in all patches together. This raises the question of how complicated could be the covariance matrix of (7.16). Even if one could reasonably assume (though not strictly) the errors of h_i and $\widetilde{h}_i^*$ to be independent, the same could not be true for the errors in the model $T_b(\mathrm{P})$; in fact, even if the gravity observations could be considered as being affected by independent measuring errors, the model is derived by solving the BVP, roughly by Stokes integration, and so it is expected to have a geographical correlation pattern. Not to be said, a correlation between the errors of $\widetilde{h}_i^*$ and T_b, as both are derived from Dg, should also be taken into account. Yet a simplification of the stochastic model, even a drastic one, would be acceptable in view of the large number of stations $\{\mathrm{P}_i\}$ that are generally available for each patch.

Nevertheless the weak point of the approach expressed by (7.16) is in the assumption that $T_b(\mathrm{P})$ is known. As a matter of fact, even the previous Earth models have always used the knowledge of low degrees coefficients of $T(\mathrm{P})$ from space geodetic observations (see for instance the paper by Rapp (1989) concerning the OSU86 model, complete up to degree and order 360). This creates models such that typically combine unbiased low degrees, derived from satellite observations, with biased gravity anomalies from ground data.

The problem will be more closely analyzed into the next sections at both local and global level.

7.3 On the Solution of the Unification Problem by a Suitable Global Model

The target of the section is to prove, by means of a careful but conservative error budget analysis, that already today we have global models that directly used in (7.15) provide us equations with errors below the 5 cm level. Since we can use several such equations for each δW_{0k}, we deem it reasonable to estimate such parameters with errors, in terms of geoid, i.e. of $\frac{\delta W_{0k}}{\gamma}$, of very few centimeters, at least in a global mean square sense. To go along this way we make beforehand two remarks.

The first is that we can free our problem from many mathematical complications if we can state a priori that all our harmonic functions can be expressed as a sum of spherical harmonics up to some finite maximum degree M; in our case M will be taken at the level of 2159, as the maximum degree of EGM2008.

This choice is justified by the following reasoning. Taking into consideration the discussion in Sect. 4.4, we start recalling the definition of full power degree variances C_n, namely

$$C_n = \sum_{m=-n}^{n} T_{nm}^2 \equiv \frac{1}{4\pi} \int \left[\sum_{m=-n}^{n} T_{nm} Y_{nm}(\sigma) \right]^2 d\sigma . \tag{7.17}$$

The plot of (7.17) for the EGM2008 model has been already displayed in Fig. 4.1. Because of their quite regular behaviour, C_n can be interpolated by some simple analytic expression. An exercise of this kind has been done by several authors with comparable results. The model that one can find in Sansò and Sideris (2013) has been computed by adapting to the empirical data the function

$$C_n = \frac{A\, q^n}{(n-1)(n-2)(n+4)(n+17)} . \tag{7.18}$$

A good matching, using only empirical values up to degree 1800, is obtained with

$$A = \left(\frac{\mu}{R}\right)^2 3.9 \cdot 10^{-5} , \quad q = 0.999443 .$$

Other authors (for example Hirt and Kuhn 2012) obtain slightly different values using all the empirical data; yet this does not change the order of magnitude of our guess. In fact adding our C_n given by (7.18) from 2160 up to 10000, we have an idea of the magnitude of the squared norm of the omitted part of T. More precisely we have the so called omission error, $\mathcal{OE}(T)$, for $M = 2159$ given by

$$\mathcal{OE}_{2160}(T) = \left\{ \frac{1}{4\pi} \int \left[\sum_{n=2160}^{+\infty} \sum_{m=-n}^{n} T_{nm} Y_{nm}(\sigma) \right]^2 d\sigma \right\}^{\frac{1}{2}} =$$
$$= \sum_{n=2160}^{+\infty} C_n \cong \sum_{n=2160}^{10000} C_n \cong 0.6 \text{ cm}\, \overline{\gamma} , \tag{7.19}$$

i.e. this omission error in terms of geoid is globally well below the centimeter value. Indeed it is clear that this does not prevent us to have a value of some centimeters in some places on the Earth surface; however this seems compatible with the target of this section.

So we shall accept the above assumption. Then we claim that, neglecting second order terms, we have

$$\mathrm{D}g + \frac{\gamma'}{\gamma}\,\delta W\bigg|_{\widetilde{S}} \cong \Delta g|_{S^*} \;, \tag{7.20}$$

with S^* the ordinary Marussi telluroid (see (4.80), (4.81)) and $\widetilde{S}$ the approximate telluroid defined by (7.14).

The relation (7.20) is proved by the following calculation

$$\begin{aligned}
\mathrm{D}g\left(\widetilde{h}^*\right) + \frac{\gamma'}{\gamma}\,\delta W\left(\widetilde{h}^*\right) &= g - \gamma\left(\widetilde{h}^*\right) + \frac{\gamma'}{\gamma}\left[W - U\left(\widetilde{h}^*\right)\right] = \\
&= g - \gamma\left(h^*\right) + \gamma\left(h^*\right) - \gamma\left(\widetilde{h}^*\right) + \frac{\gamma'}{\gamma}\left[U\left(h^*\right) - U\left(\widetilde{h}^*\right)\right] = \\
&= \Delta g\left(h^*\right) + \gamma'\left(h^* - \widetilde{h}^*\right) + \frac{\gamma'}{\gamma}\left[-\gamma\left(h^* - \widetilde{h}^*\right)\right] = \Delta g\left(h^*\right) \;.
\end{aligned}$$

We would like to acknowledge that this complies with a personal communication of T. Krarup to one of the authors.

A consequence of this remark is that, since the solution of the GBVP is unique, solving such a problem with known term Δg on S^* or with $\mathrm{D}g + \frac{\gamma'}{\gamma}\mathrm{D}W$ on $\widetilde{S}$ should give the same result in the linear approximation. Concisely, introducing the two solver operators $\mathcal{S}^*$ and $\widetilde{\mathcal{S}}$, the former referring to the GBVP with S^* as boundary, the latter to the same problem with $\widetilde{S}$ as boundary, we can claim that

$$T \cong \mathcal{S}^*\left(\Delta g\right) \cong \widetilde{\mathcal{S}}\left(\mathrm{D}g + \frac{\gamma'}{\gamma}\,\mathrm{D}W\right) \;. \tag{7.21}$$

Now we are ready to introduce our simple minded global model $\widetilde{T}$. We started by observing that we have available satellite-only models combining data from satellite geodesy of different missions, particularly the models derived by the three gravimetric/grodiometric missions CHAMP, GRACE and GOCE (Reigber et al. 2004; Tapley et al. 2004; Pail et al. 2011). Specifically we shall refer to the GOCO model T^{G} (Pail et al. 2010; Mayer-Gürr et al. 2015) up to degree and order 200, a level at which the cumulated error for the estimate of the coefficients becomes larger than the magnitude of the coefficients themselves, expressed by their degree variances C_n. So up to degree 200 we follow T^{G}, knowing that

$$T^{\mathrm{G}} = T_{200} + \varepsilon^{\mathrm{G}} \tag{7.22}$$

with

$$\frac{1}{\overline{\gamma}}\,\sigma\left(T^{\mathrm{G}}\right) = \frac{1}{\overline{\gamma}}\,\sigma\left(\varepsilon^{\mathrm{G}}\right) \cong 2\ \mathrm{cm}\;, \tag{7.23}$$

as it results from the estimates of the degree standard deviations provided with the model. The quantity $\sigma\left(\varepsilon^{\mathrm{G}}\right)$ in (7.23) is called the commission error $\mathcal{CE}\left(\varepsilon^{\mathrm{G}}\right)$ (see (4.40)).

We strengthen again that if we introduce the projection operator $\mathcal{P}_L$ that cuts every harmonic function at the degree L, i.e. for $M > L$

$$\mathcal{P}_L\left(\sum_{n=2}^{M} u_{nm}\left(\frac{R}{r}\right)^{n+1} Y_{nm}(\sigma)\right) = \sum_{n=2}^{L} u_{nm}\left(\frac{R}{r}\right)^{n+1} Y_{nm}(\sigma) , \tag{7.24}$$

then indeed, with $L = 200$, we have

$$\mathcal{P}_L T^{\mathrm{G}} \equiv T^{\mathrm{G}} , \tag{7.25}$$

so that (7.22) more precisely reads

$$T^{\mathrm{G}} = \mathcal{P}_L T + \varepsilon^{\mathrm{G}} . \tag{7.26}$$

Moreover the explicit interpretation of (7.23) is

$$\frac{1}{\overline{\gamma}}\sigma\left(\varepsilon^{\mathrm{G}}\right) = \frac{1}{\overline{\gamma}}\left[\mathrm{E}\left(\frac{1}{4\pi}\int\left(\varepsilon^{\mathrm{G}}\right)^2 \mathrm{d}\sigma\right)\right]^{\frac{1}{2}} ,$$

where the expectation E is taken on the stochastic structure of ε^{G}.

We assume that the information contained in T^{G} is better than the corresponding information on the low degrees contained in the EGM2008 model T^{E}. On the contrary, for degrees higher than 200 the only global information (in reality up to degree 2159) we have is contained in T^{E}, so we will take it as it is. Therefore we propose to create a kind of "Frankenstein model" according to

$$\widetilde{T} = T^{\mathrm{G}} + (I - \mathcal{P}_L)\, T^{\mathrm{E}} . \tag{7.27}$$

We note however that T^{E} has been computed from ground data, at least in the range of degrees higher than 200, and so it is affected by a bias because it could only be computed from the observations $\mathrm{D}g_0$ on the approximate telluroid $\widetilde{S}$. In other words

$$(I - \mathcal{P}_L)\, T^{\mathrm{E}} = (I - \mathcal{P}_L)\, \widetilde{\mathcal{S}}\left(\mathrm{D}g_0\right) . \tag{7.28}$$

Indeed $\mathrm{D}g_0$ is affected by some noise ε^g that propagates to the solution

$$\widetilde{\mathcal{S}}\left(\mathrm{D}g_0\right) = \widetilde{\mathcal{S}}\left(\mathrm{D}g + \varepsilon^g\right) = \widetilde{\mathcal{S}}\left(\mathrm{D}g\right) + \varepsilon^{\mathrm{E}} . \tag{7.29}$$

This ε^{E} is what in literature is called the commission error $\mathcal{CE}\left(\varepsilon^{\mathrm{E}}\right)$ of the model, and it is clear from our reasoning that ε^{E} has a maximum degree equal to 2159 too.

With such specifications, (7.27) reads

$$\widetilde{T} = \mathcal{P}_L T + \varepsilon^{\mathrm{G}} + (I - \mathcal{P}_L)\, \widetilde{\mathcal{S}}\, (\mathrm{D}g) + (I - \mathcal{P}_L)\, \varepsilon^{\mathrm{E}} \,. \tag{7.30}$$

Also for ε^{E} we have estimates that come together with the model T^{E}; the number cumulating all the errors between degree 200 and 2159 is

$$\mathcal{CE}\left(\varepsilon^{\mathrm{E}}\right) = \mathrm{E}\left[\frac{1}{4\pi}\int \left[(I - \mathcal{P}_L)\, \varepsilon^{\mathrm{E}}\right]^2 \mathrm{d}\sigma\right]^{\frac{1}{2}} = 3.6\ \mathrm{cm}\,. \tag{7.31}$$

If we write the analogous of (7.30) for T, also taking into account (7.21), we see that

$$\begin{aligned} T &= \mathcal{P}_L T + (I - \mathcal{P}_L)\, \mathcal{S}^* (\Delta g) = \\ &= \mathcal{P}_L T + (I - \mathcal{P}_L)\, \widetilde{\mathcal{S}}\, (\mathrm{D}g) + (I - \mathcal{P}_L)\, \widetilde{\mathcal{S}} \left(\frac{\gamma'}{\gamma}\, \delta W\right) . \end{aligned} \tag{7.32}$$

Comparing (7.32) and (7.30), we find the total estimation error of $\widetilde{T}$, namely

$$\widetilde{T} - T = \varepsilon^{\mathrm{G}} + (I - \mathcal{P}_L)\, \varepsilon^{\mathrm{E}} - (I - \mathcal{P}_L)\, \widetilde{\mathcal{S}} \left(\frac{\gamma'}{\gamma}\, \delta W\right) . \tag{7.33}$$

If we can suppose that ε^{G} and ε^{E} have zero average, the same is not justified for $(I - \mathcal{P}_L)\, \widetilde{\mathcal{S}} \left(\frac{\gamma'}{\gamma}\, \delta W\right)$, which then assumes the meaning of the bias of $\widetilde{T} - T$, i.e.

$$b\,(\mathrm{P}) = \mathrm{E}\left\{\widetilde{T} - T\right\} = -\,(I - \mathcal{P}_L)\, \widetilde{\mathcal{S}} \left(\frac{\gamma'}{\gamma}\, \delta W\right) . \tag{7.34}$$

The construction of our error budget then continues with a majorization of the mean quadratic value of $b\,(\mathrm{P})$ over the unit sphere.

Now consider that $\widetilde{\mathcal{S}}$, the BVP solver, is as a matter of fact a combination of some kind of regularized downward continuation to the Earth ellipsoid and then a solution by quadrature with spherical harmonics (Sansò and Sideris 2013, Part II, Chap. 6). In any event, due to the smallness of the function

$$\frac{\gamma'}{\gamma}\, \delta W \cong -\frac{2}{r}\, \delta W\,, \tag{7.35}$$

(remember that $\mathcal{O}\left(\frac{\delta W}{\overline{\gamma}}\right) \cong 2$ m), we can approximate $\widetilde{\mathcal{S}}$ as applied to (7.35) by a simple spherical solver, namely the Stokes integral, which certainly constitutes the "large part" of $\widetilde{\mathcal{S}}$. So we can write (see (4.100))

$$b = -(I - \mathcal{P}_L)\,\widetilde{\mathcal{S}}\left(\frac{\gamma'}{\gamma}\,\delta W\right) \cong \frac{2}{R_0}\sum_{n=L+1}^{M}\sum_{m=-n}^{n}\frac{R_0}{n-1}\,\delta W_{nm}\,Y_{nm}(\sigma) \tag{7.36}$$

with R_0 the mean Earth radius. From (7.36) we then derive

$$\begin{aligned}\|b\|^2_{L^2_\sigma} &= \left\|(I - \mathcal{P}_L)\,\widetilde{\mathcal{S}}\left(\frac{\gamma'}{\gamma}\,\delta W\right)\right\|^2_{L^2_\sigma} = 4\sum_{n=L+1}^{M}\sum_{m=-n}^{n}\frac{\delta W^2_{nm}}{(n-1)^2} \le \\ &\le \frac{4}{L^2}\sum_{n=L+1}^{M}\sum_{m=-n}^{n}\delta W^2_{nm} = \frac{4}{L^2}\,\|(I-\mathcal{P}_L)\,\delta W\|^2_{L^2_\sigma} < \\ &< \frac{4}{L^2}\,\|\delta W\|^2_{L^2_\sigma}\ .\end{aligned} \tag{7.37}$$

Now we observe that, owing to its definition (7.4), δW^2 is given by

$$\delta W^2\,(\mathrm{P}) = \sum_{k=1}^{K}\delta W^2_{0k}\,\chi_k\,(\mathrm{P})\ ,$$

so that

$$\|\delta W\,(\mathrm{P})\|^2_{L^2_\sigma} = \frac{1}{4\pi}\int\sum_{k=1}^{K}\delta W^2_{0k}\,\chi_k\,(\mathrm{P})\,\mathrm{d}\sigma = \sum_{k=1}^{K}\delta W^2_{0k}\,\frac{|A_k|}{4\pi}\ , \tag{7.38}$$

where we have designated by $|A_k|$ the area of the patch A_k, projected on the unit sphere. As we see, (7.38) is a kind of weighted average of the δW^2_{0k} and, since $\max\left|\dfrac{\delta W_{0k}}{\overline{\gamma}}\right| \le 2$ m, we could reasonably hypothesize that

$$\frac{1}{\overline{\gamma}}\left\{\sum_{k=1}^{K}\delta W^2_{0k}\,\frac{|A_k|}{4\pi}\right\}^{\frac{1}{2}} \le 1\ \mathrm{m}\ . \tag{7.39}$$

Using (7.39) in (7.37), we receive

$$\frac{1}{\overline{\gamma}}\,\|b\|_{L^2_\sigma} < \frac{2}{200}\cdot 1\ \mathrm{m} = 1\ \mathrm{cm}\ . \tag{7.40}$$

Putting (7.23), (7.32) and (7.40) together, we formulate the following error budget

$$\begin{aligned}\frac{1}{\overline{\gamma}}\left\{\mathrm{E}\left[\|\widetilde{T} - T\|^2_{L^2_\sigma}\right]\right\}^{\frac{1}{2}} &= \frac{1}{\overline{\gamma}}\left\{\mathcal{CE}^2\left(\varepsilon^{\mathrm{G}}\right) + \mathcal{CE}^2\left(\varepsilon^{\mathrm{E}}\right) + \|b\|^2_{L^2_\sigma}\right\}^{\frac{1}{2}} \le \\ &\le \{4 + 12.96 + 1\}^{\frac{1}{2}}\ \mathrm{cm} = 4.24\ \mathrm{cm}\ .\end{aligned} \tag{7.41}$$

Let us remark that, if instead of (7.39) we had taken the upper limit of 2 m, then (7.41) would rise to 4.58 cm, which is not a very different number.

Let us further observe that certainly our analysis here is not very refined and in particular the model $\widetilde{T}$ on which the error budget has been constructed is not the optimal that one could calculate. Optimal solutions of the combination of satellite and existing global models can be found in literature (see for example Pavlis et al. 2012, 2013; Reguzzoni and Sansò 2012; Sansò and Sideris 2013, Part II, Chap. 6; Gilardoni et al. 2016).

On the other hand, we promised a conservative analysis that has generated the figure of 5 cm to majorize our global error; so we are confident that this is a reliable upper bound. Since the large part of the index (7.41) is due to $\mathcal{CE}\left(\varepsilon^{\mathrm{E}}\right)$, we know that this index has a great geographic variability, reaching the level of 30–40 cm in the Himalayas and in the Andes when ε^{E} includes also the first 200 degrees. However this is not the case in most areas of the globe and we can expect that a figure between 5 and 10 cm could be respected by the error in the stations chosen to construct the system (7.16). Therefore a first proposal is to use $\widetilde{T}$ (or a better model) in (7.16), so that we can write observation equations patch by patch and, hopefully, by averaging we can resort to an estimate of $\dfrac{\delta W_{0k}}{\gamma}$ with a few centimeters error.

A more refined proposal is to use the model $\widetilde{T}$ to arrive at a system of equations similar to (7.16); however we have now to pay attention to split the degrees below and above 200, as discussed in this section. In this case, from Eqs. (7.15), (7.33) and (7.36) we could write

$$\begin{aligned} h_i - \widetilde{h}_i^* &= \frac{\widetilde{T}}{\gamma} + \frac{T-\widetilde{T}}{\gamma} + \frac{\delta W_{0k}}{\gamma} = \\ &= \frac{\widetilde{T}}{\gamma} - \frac{1}{\gamma}(I - \mathcal{P}_L)\,\widetilde{\mathcal{S}}\left(\frac{2}{r}\,\delta W\right) + \frac{\delta W_{0k}}{\gamma} = \\ &= \frac{\widetilde{T}}{\gamma} - \frac{2}{\gamma}\sum_{n=L+1}^{M}\sum_{m=-n}^{n}\frac{\delta W_{nm}}{n-1}\,Y_{nm}\left(\sigma_{\mathrm{P}_i}\right) + \frac{\delta W_{0k}}{\gamma}\,; \end{aligned} \tag{7.42}$$

note that in (7.42) only the deterministic terms are reported, leaving the stochastic errors aside.

Now considering that

$$\delta W_{nm} = \sum_{j=1}^{K}\delta W_{0j}\,\langle \chi^j, Y_{nm}\rangle = \sum_{j=1}^{K}\delta W_{0j}\,\chi^j_{nm}\,,$$

Eq. (7.42) can be rewritten in the form

$$h_i - \widetilde{h}_i^* = \frac{\widetilde{T}\,(\mathrm{P}_i)}{\gamma} - \frac{2}{\gamma}\sum_{j=1}^{K}\delta W_{0j}\left[\sum_{n=L+1}^{M}\sum_{m=-n}^{n}\frac{\chi^j_{nm}}{n-1}\,Y_{nm}\left(\sigma_{\mathrm{P}_i}\right)\right] + \frac{\delta W_{0k}}{\gamma}\,, \tag{7.43}$$

where the unknown parameters $\{\delta W_{0k}\}$ appear explicitly and all the other terms are either observed or computed.

When all the quantities h_i, $\widetilde{h}_i^*$ and $\widetilde{T}(\mathrm{P}_i)$ are derived from observations, the Eq. (7.43) should be complemented with the proper error terms; if we assume that the errors in h_i and $\widetilde{h}_i^*$ are in the range of millimeters, and therefore negligible, and recalling (7.33), we can write

$$h_i - \widetilde{h}_i^* = \frac{\widetilde{T}(\mathrm{P}_i)}{\gamma} - \frac{2}{\gamma}\sum_{j=1}^{K}\delta W_{0j}\left[\sum_{n=L+1}^{M}\sum_{m=-n}^{n}\frac{\chi_{nm}^{j}}{n-1}Y_{nm}\left(\sigma_{\mathrm{P}_i}\right)\right] + \\ + \frac{\delta W_{0k}}{\gamma} + \varepsilon^{\mathrm{G}} + (I - \mathcal{P}_L)\,\varepsilon^{\mathrm{E}}\,. \tag{7.44}$$

7.4 On Local Solutions of the Height Datum Problem

We have already mentioned in the previous section that, when we have available a good model of the anomalous potential, like our $\widetilde{T}$ or better, we can safely substitute it in observation equations of the shape (7.15). This implies neglecting the bias term (7.36), which has been estimated to globally produce (cfr. (7.38)) a mean square error between 1 and 2 cm, and to accept a stochastic error, $\varepsilon^{\mathrm{G}} + (I - \mathcal{P}_L)\,\varepsilon^{\mathrm{E}}$, with an overall magnitude of the order of 4 cm. Including all the effects into the observation equation, we arrive at formula (7.44).

However two aspects limit this global approach to the determination of the height datum, i.e. of the biases $\{\delta W_{0k}\}$, namely that in oceanic areas we have observations for h_i ($\widetilde{h}_i^* = 0$ in this case) but this dataset is strongly correlated and the covariance structure of the error is not really known; moreover biases and stochastic errors can have a strong geographic signature which could deviate the estimates of $\frac{\delta W_{0k}}{\gamma}$, by one or more decimeters, at least for particular areas.

This is ultimately due to the fact that in such areas $\widetilde{T}$ is not a sufficient approximation to T; however we know that, apart from biases, we are able to compute a better estimate of T, for instance by using a local collocation solution

$$\widehat{T}_{\mathrm{loc}} = \widetilde{T} + T_{\mathrm{res}}\,, \tag{7.45}$$

for which a typical error-figure in terms of height anomaly could be 1–2 cm. We will call $\varepsilon_{\mathrm{res}}$ the error associated to the estimated residual potential T_{res}.

We want to examine whether and how we could take advantage of this improved knowledge to estimate one of the biases for a specific area. In this case we have to return to (7.44) and use $\widehat{T}_{\mathrm{loc}}$ instead of $\widetilde{T}$ and $\varepsilon_{\mathrm{res}}$ instead of $\varepsilon^{\mathrm{G}} + (I - \mathcal{P}_L)\,\varepsilon^{\mathrm{E}}$, thus arriving at an observation equation that we rewrite in the form

$$h_i - \widetilde{h}_i^* = \frac{\widehat{T}_{\mathrm{loc}}(\mathrm{P}_i)}{\gamma} - \frac{2}{\gamma}\sum_{j=1, j\neq k}^{K} \delta W_{0j}\left[\sum_{n=L+1}^{M}\sum_{m=-n}^{n}\frac{\chi_{nm}^j}{n-1}Y_{nm}\left(\sigma_{\mathrm{P}_i}\right)\right] +$$
$$+ \frac{\delta W_{0k}}{\gamma}\left\{1 - 2\left[\sum_{n=L+1}^{M}\sum_{m=-n}^{n}\frac{\chi_{nm}^k}{n-1}Y_{nm}\left(\sigma_{\mathrm{P}_i}\right)\right]\right\} + \varepsilon_{\mathrm{res},i} \quad \mathrm{P}_i \in A_k \ . \tag{7.46}$$

Let us consider one element of the sum in the second term of the right hand side of (7.46), namely

$$\frac{2}{\gamma}\delta W_{0j}\left[\sum_{n=L+1}^{M}\sum_{m=-n}^{n}\frac{\chi_{nm}^j}{n-1}Y_{nm}\left(\sigma_{\mathrm{P}_i}\right)\right] \cong$$
$$\cong \frac{1}{\gamma}\left(I - \mathcal{P}_L\right)\widetilde{\mathcal{S}}\left(\frac{2\,\delta W_{0j}}{R}\chi^j\right) \cong$$
$$\cong \frac{\delta W_{0j}}{\gamma}\frac{2}{4\pi}\int_{A_j}\sum_{n=L+1}^{M}\frac{2n+1}{n-1}P_n\left(\cos\psi_{\mathrm{P}_i\mathrm{Q}}\right)d\sigma_{\mathrm{Q}} \quad (j \neq k) \ . \tag{7.47}$$

As we see such a term represents the influence of the bias δW_{0j} of the zone A_j in the area A_k; when the two are well separated, it is known that the influence function

$$F_j(\mathrm{P}) = \frac{1}{2\pi}\int_{A_j}\sum_{n=L+1}^{M}\frac{2n+1}{n-1}P_n\left(\cos\psi_{\mathrm{P}_i\mathrm{Q}}\right)d\sigma_{\mathrm{Q}} \ , \tag{7.48}$$

i.e. the integral on A_j of the truncated Stokes function, becomes quite small. However, if we could simply ignore $F_j(\mathrm{P})$, even when A_j is a neighbour of A_k, then we could delete the second term in (7.46), which at this point would become an observation equation for δW_{0k} only, i.e. we would have the possibility of a local determination of the bias.

Note that what we need now is a pointwise estimate for $\left|F_j(\mathrm{P})\right|$ and not the global mean square estimate that has already been found in the previous section. Unfortunately we do not have a strict proof, but only a guess based on the following example.

Example Assume A_j is just a spherical cap C_Δ of radius Δ, then we shall prove that the following approximate majorization holds

$$|F(\mathrm{P})| \lessapprox \frac{2}{\pi}\frac{1}{L+1} \tag{7.49}$$

when P is on the boundary of C_Δ, irrespectively of the value of Δ.

If we take the origin of the spherical coordinates at the centre of C_Δ, from (4.45) and using the summation theorem, we have

$$
\begin{aligned}
F(\mathrm{P}) &= \frac{1}{2\pi} \sum_{n=L+1}^{M} \sum_{m=-n}^{n} \frac{Y_{nm}(\mathrm{P})}{n-1} \int_{C_\Delta} Y_{nm}(\mathrm{Q})\, \mathrm{d}\sigma_{\mathrm{Q}} = \\
&= \frac{1}{2\pi} \sum_{n=L+1}^{M} \frac{Y_{n0}(\mathrm{P})}{n-1} \int_{C_\Delta} Y_{n0}(\mathrm{Q})\, \mathrm{d}\sigma_{\mathrm{Q}} = \\
&= \sum_{n=L+1}^{M} \frac{(2n+1)\, P_n(\cos\theta_{\mathrm{P}})}{n-1} \int_0^{\Delta} P_n(\cos\theta) \sin\theta\, \mathrm{d}\theta\,.
\end{aligned} \tag{7.50}
$$

Since

$$
\begin{aligned}
(2n+1) \int_0^{\Delta} P_n(\cos\theta) \sin\theta\, \mathrm{d}\theta &= \int_{\cos\Delta}^{1} (2n+1)\, P_n(t)\, \mathrm{d}t = \\
= \int_{\cos\Delta}^{1} \left[P'_{n+1}(t) - P'_{n-1}(t)\right] \mathrm{d}t &= P_{n-1}(\cos\Delta) - P_{n+1}(\cos\Delta)\ ,
\end{aligned} \tag{7.51}
$$

Eq. (7.50) becomes

$$
F(\mathrm{P}) = \sum_{n=L+1}^{M} \frac{P_n(\cos\theta_{\mathrm{P}})}{n-1} \left[P_{n-1}(\cos\Delta) - P_{n+1}(\cos\Delta)\right]. \tag{7.52}
$$

Now we apply a famous asymptotic expression for the Legendre polynomials (Abramowitz and Stegun 1964) claiming that

$$
P_n(\cos\theta) = \sqrt{\frac{2}{\pi \sin\theta \cdot n}} \cos\left[\left(n+\frac{1}{2}\right)\theta - \frac{\pi}{4}\right] + \mathcal{O}\left(n^{-\frac{3}{2}}\right). \tag{7.53}
$$

In particular (7.53) holds for

$$
\theta > \frac{3\pi}{4n+2}\ ; \tag{7.54}
$$

since we have in mind that $n > 200$ and Δ is at least 2° or (much) more, the condition (7.54) is met.

So we proceed noting that

$$
\begin{aligned}
&P_{n-1}(\cos\Delta) - P_{n+1}(\cos\Delta) \cong \\
&\cong \sqrt{\frac{2}{\pi\sin\Delta}} \left\{ \frac{\cos\left[\left(n+\frac{1}{2}-1\right)\Delta - \frac{\pi}{4}\right]}{\sqrt{n-1}} - \frac{\cos\left[\left(n+\frac{1}{2}+1\right)\Delta - \frac{\pi}{4}\right]}{\sqrt{n+1}} \right\}.
\end{aligned}
$$

Since

$$\frac{1}{\sqrt{n \pm 1}} - \frac{1}{\sqrt{n}} = \mathcal{O}\left(\frac{1}{n^{\frac{3}{2}}}\right) ,$$

the above asymptotic relation can be written as

$$\begin{aligned}
&P_{n-1}(\cos\Delta) - P_{n+1}(\cos\Delta) \cong \\
&\cong \sqrt{\frac{2}{\pi \sin\Delta \cdot n}} \left\{\cos\left[\left(n+\frac{1}{2}-1\right)\Delta - \frac{\pi}{4}\right] - \cos\left[\left(n+\frac{1}{2}+1\right)\Delta - \frac{\pi}{4}\right]\right\} = \\
&= \frac{2\sqrt{2}}{\pi}\sqrt{\frac{\sin\Delta}{n}} \sin\left[\left(n+\frac{1}{2}\right)\Delta - \frac{\pi}{4}\right] .
\end{aligned}$$

So returning to (7.52) and applying (7.53) to $P_n(\cos\theta_{\mathrm{P}})$ too, we find

$$F(\mathrm{P}) = \frac{4}{\pi}\sqrt{\frac{\sin\Delta}{\sin\theta_{\mathrm{P}}}} \sum_{n=L+1}^{M} \frac{\cos\left[\left(n+\frac{1}{2}\right)\theta_{\mathrm{P}} - \frac{\pi}{4}\right] \sin\left[\left(n+\frac{1}{2}\right)\Delta - \frac{\pi}{4}\right]}{n(n-1)} . \tag{7.55}$$

As soon as we put P on the boundary of C_Δ, i.e. we take $\theta_{\mathrm{P}} = \Delta$, we get from (7.55)

$$\begin{aligned}
|F(\mathrm{P})| &\cong \frac{2}{\pi}\left|\sum_{n=L+1}^{M} \frac{\sin\left[(2n+1)\Delta - \frac{\pi}{2}\right]}{n(n-1)}\right| \lessapprox \\
&\lessapprox \frac{2}{\pi}\sum_{n=L+1}^{+\infty} \frac{1}{n^2} \cong \frac{2}{\pi}\frac{1}{L+1} ,
\end{aligned}$$

and (7.49) is proved.

With this example we see that, at least when A_j is a spherical cap and $L = 200$, the influence of the bias δW_{0j} at its boundary is

$$\left|\frac{\delta W_{0j}}{\gamma} F(\mathrm{P})\right| \leq \left|\frac{\delta W_{0j}}{\gamma}\right| 3.2 \cdot 10^{-3} ,$$

namely well below the 1 cm level, even when $\dfrac{\delta W_{0j}}{\gamma} = 2$ m. Indeed when $\theta_{\mathrm{P}} > \Delta$, we expect that $F(\mathrm{P})$ is even smaller, as shown in Fig. 7.1 when $\Delta = 5°$.

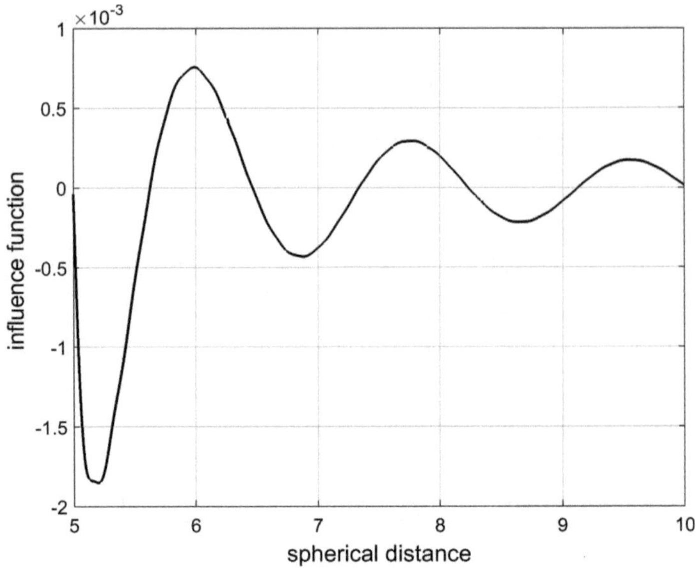

Fig. 7.1 Influence function $F(\theta_P)$ for $5° \leq \theta_P \leq 10°$, in the case of a spherical cap C_Δ with $\Delta = 5°, L = 200$

Based on the guess supported by the above example, we propose that a local bias δW_{0k} is estimated from the set of observation equations

$$h_i - \widetilde{h}_i^* = \frac{\widehat{T}_{\text{loc}}(\text{P}_i)}{\gamma} + \frac{\delta W_{0k}}{\gamma}\left[1 - F_k(\text{P}_i)\right] + \varepsilon_{\text{res},i}\,, \tag{7.56}$$

where P_i are all the points in the area A_k where both h_i and $\widetilde{h}_i^*$ are available.

We close the section by observing that indeed we could have a situation where several $\{\delta W_{0k}\}$ can be estimated together, although they refer to some areas that do not cover the whole sphere, with an obvious modification of the above discussion. We underline however that in this case it is better that the local estimate of the potential $\widehat{T}_{\text{loc}}$ is computed for the above areas together, because only in this case we shall have a consistent covariance matrix for ε_{res} (Reguzzoni and Venuti 2018).

7.5 An Example: The Italian Case

In this paragraph, the local solution of the height datum problem discussed in Sect. 7.4 is applied to the Italian case study. A similar computation has been applied as well to the determination of the geoid bias in Spain (Reguzzoni et al. 2018).

As a matter of facts, Italy has three different height systems based on three different reference tide gauges. The reference tide gauge for the mainland is in Genoa, while heights in Sicily are referred to the Catania tide gauge and those of Sardinia to Cagliari. Due to the different dynamic ocean topography in these three reference stations, inconsistencies at the decimetre level among heights in Italy mainland, Sicily and Sardinia are expected.

The equation to be used in estimating the local biases is (7.56) which can be further simplified for the present computation. In fact, in the Italian case presented here, it can be numerically proved that even considering the complete Eq. (7.16) accounting for the global unification, the term

$$\frac{1}{\gamma}\sum_{j=1}^{K}\delta W_{0j}\,F_j\,(\mathrm{P}_i)$$

is smaller than 1 mm. Thus, a fortiori, the corresponding local term in (7.56) can be disregarded.

So, the equation that will be used in the computation is

$$h_i - \widetilde{h}_i^* = \frac{\widehat{T}_{\mathrm{loc}}\,(\mathrm{P}_i)}{\gamma} + \frac{\delta W_{0k}}{\gamma} + \varepsilon_{\mathrm{res},i} \tag{7.57}$$

that can be rewritten as

$$\widetilde{\zeta}_k\,(\mathrm{P}_i) = \frac{\widehat{T}_{\mathrm{loc}}\,(\mathrm{P}_i)}{\gamma} + b_k + \varepsilon_{\mathrm{res},i} \tag{7.58}$$

where $\widetilde{\zeta}_k\,(\mathrm{P}_i)$ are the biased height anomalies in the k-th area and b_k the bias to be estimated on the same area.

It can be further assumed that $\widehat{T}_{\mathrm{loc}}$ is estimated as

$$\widehat{T}_{\mathrm{loc}}\,(\mathrm{P}_i) = T^L\,(\mathrm{P}_i) + T^H\,(\mathrm{P}_i) \tag{7.59}$$

where T^L is the prediction of the anomalous potential at point P_i coming from a satellite gravity model to degree L and T^H is the prediction derived from a high degree model, like e.g. EGM2008, from degree $L + 1$ to degree H. Although by considering T^H we reintroduce biases through ground gravity data, it can be proved that the impact on the solution is of the order of some millimetres (Gatti et al. 2013). Thus, one can say that a feasible solution for the estimate of b_k can be obtained by the observation equation

$$\widetilde{\zeta}_k\,(\mathrm{P}_i) = \frac{T^L\,(\mathrm{P}_i) + T^H\,(\mathrm{P}_i)}{\gamma} + b_k + \varepsilon_{\mathrm{res},i} \tag{7.60}$$

By separating the observations and the unknowns to be estimated, one gets

$$\widetilde{\zeta}_k\left(\mathrm{P}_i\right)-\frac{T^L\left(\mathrm{P}_i\right)+T^H\left(\mathrm{P}_i\right)}{\gamma}+\varepsilon_{\mathrm{res},i}=b_k\ . \tag{7.61}$$

Now, if one considers N points in the K regions, with $N \geq K$, a linear system of N equations and K unknowns can be solved by least squares adjustment, once the observation error covariance matrix of ε is defined. This matrix has to account for the dispersion of the errors in the ellipsoidal heights derived from GNSS through the covariance matrix C_h, the errors in the normal heights derived from levelling and gravity measurements through $C_{\widetilde{h}^*}$, the commission errors of the satellite-only gravity model up to the degree L through C_{T^L}, and those in the high resolution model from degree $L+1$ up to degree H through C_{T^H}. Thus, assuming the above described errors independent from one another, the proper covariance structure to be used in the adjustment procedure is

$$C_\varepsilon=C_\zeta+C_{T^L}+C_{T^H}=C_h+C_{\widetilde{h}^*}+C_{T^L}+C_{T^H}\ . \tag{7.62}$$

In the Italian test case the least square problem is set by considering 1,068 points with known GNSS ellipsoidal heights and levelling derived heights. Among them, 43 points are in Sicily, 48 in Sardinia and the remaining 977 in the Italian mainland. The heights derived from levelling measurements were obtained by a least squares adjustment of the observations without any correction accounting for gravity effects (Betti et al. 2016). GNSS heights are referred to the ETRF2000 reference frame, epoch 2008.0.

Hence, it must be underlined that in Eq. (7.61) biased geoid undulations $\widetilde{N}_k\left(\mathrm{P}_i\right)$ are used (which, as said, are further biased since no gravity corrections have been applied).

The models components that have been considered in order to evaluate the T^L and the T^H terms are the GOCO-03S satellite gravity only model (Mayer-Gürr et al. 2012) and the EGM2008 global geopotential model. The GOC0-03S model basically combines the ITG-Bonn GRACE solution with the time-wise GOCE one (release R3, that is the third solution based on 1 year and a half GOCE data). The coefficients are available at the website of the International Center for Global Earth Models (ICGEM). Moreover, the GOC0-03S order-wise block diagonal error covariance matrix has been considered in the computation, which practically bears the same information as the full error covariance matrix (Gerlach and Fecher 2012). As for EGM2008 spherical harmonic coefficients, the error coefficient variances and a global grid of local geoid error variances are available. Consistently with GNSS data, the coefficients of the two global models are tide-free, while the levelling data are referred to the mean sea level at the three tide gauges of Genoa, Catania and Cagliari.

Before computing the left hand side of Eq. (7.61), reference frame transformations have to be considered.

The different coordinates have been referred to the most recent frame of the GOCE model. The Italian GNSS data are given in ETRF2000, epoch 2008.0, while GOCE data are in ITRF2008, with an unspecified epoch between 2010 and 2011. Transformations from ETRF2000-2008.0 to ITRF2008-2010/2011 can be performed in three steps. The EUREF transformation parameters have been applied from ETRF2000-2008 to ITRF2000-2008 and then the IERS transformation parameters have been used from ITRF2000-2008 to ITRF2008-2008. Finally, the ITRF2008 coordinates have been updated to epoch 2010/2011 using the mean velocity of a subset of Italian GNSS permanent stations (velocities published by IERS). To this aim, the GNSS permanent stations of Medicina, Genoa, Torino I, Cagliari, Matera, Padova and Perugia have been taken into account.

These transformations accounted for a displacement in the horizontal coordinates of about 50 cm and a 1 cm change in their heights. It can be proved that the impact of these shifts is negligible in terms of the bias estimation (Barzaghi et al. 2016). Similar transformations were not applied to EGM2008 since its reference time is not available.

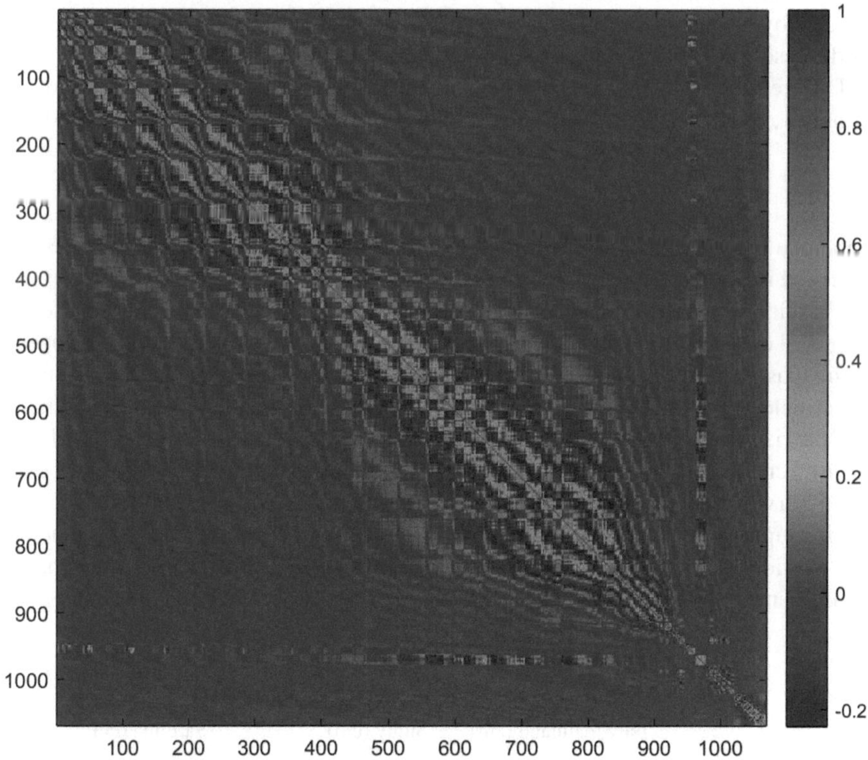

Fig. 7.2 The correlation matrix of ε with $L = 250$

Another key point in setting up the least squares problem in (7.61) is the definition of the stochastic model of the observations. This stochastic model, represented in (7.62), can be evaluated from the available error models. The set of differences between GNSS and levelling heights are assumed to be as an uncorrelated noise, so that it can be set

$$C_\varsigma = \sigma_\varsigma^2 I \tag{7.63}$$

where I is the identity matrix. The error covariance matrix C_{T^L} of the set of potential values T^L predicted in the GNSS-leveling points from GOC0-03S is obtained by propagation from the given order-wise block diagonal error covariance matrix. The covariance matrix C_{T^H} of the set of potential values T^H computed at the same points from EGM2008, is obtained by propagation from the coefficient error variances properly rescaled accordingly to the geographical map of local geoid errors (Gilardoni et al. 2013). The resulting error correlation matrix, with $\sigma_\varsigma^2 = 1$ cm and $L = 250$, is plotted in Fig. 7.2.

Based on this covariance structure, the error in the estimated biases can be computed as a function of the degree L. In the Italian case study, it can be shown that the errors in the estimated biases of Italy mainland, Sicily and Sardinia are not strongly affected by the choice of L (Barzaghi et al. 2016) so that $L = 250$, the full GOC0-03S model resolution, has been selected in the computation.

Different biases estimates have been then computed using different values of σ_ς, namely 1, 5, 10 and 12 cm. The least squares estimate satisfying the null hypothesis test

$$H_0 : \sigma_0^2 = 1$$

is the one based on $\sigma_\varsigma = 12$ cm, which gives the values for the estimated biases that are listed in Table 7.1.

This first result is based on some quite strong simplifications and is hence affected by model errors. Particularly, the use of $\widetilde{N}_k$ (P$_i$), the biased geoid undulation, instead of the biased height anomaly $\widetilde{\zeta}_k$ (P$_i$), can induce distortions in the estimated biases. Nevertheless, the difference between the biases of Italy mainland and Sicily, that is 9.82 cm, is significantly close to the values reported by Istituto Geografico Militare, i.e. 14.1 cm. It is to be underlined that this value has been independently estimated using surveying techniques based on trigonometric levelling trough the Messina Strait coupled with spirit levelling in Sicily and Calabria to form a close loop across the two sides of the strait. So, despite the use of somehow improper data, acceptable results can be obtained by the devised least squares adjustment procedure. Thus, one

Table 7.1 The estimated biases with $\sigma_\varsigma = 12$ cm

	Italy mainland (cm)	Sicily (cm)	Sardinia (cm)
b_k	77.22	67.40	97.90
σ_{b_k}	0.52	2.57	2.72

comes to a confirmation that the proposed method is effective in estimating the local biases and can be applied for solving the problem of the height systems unification at local level.

References

Abramowitz M., Stegun I.A. (1964). *Handbook of mathematical functions with formulas, graphs, and mathematical tables*. Dover Publications, New York.

Barzaghi R., Carrion D., Reguzzoni M., Venuti G. (2016). *A feasibility study on the unification of the Italian height systems using GNSS-leveling data and global satellite gravity models*. In: Rizos C., Willis P. (eds), *IAG 150 Years*, International Association of Geodesy Symposia, 143:281–288, Springer, Cham.

Betti B., Carrion D., Sacerdote F., Venuti G. (2016). *The observation equation of spirit leveling in Molodensky's context*. In: Sneeuw N., Novàk P., Crespi M., Sansò F. (eds), VIII Hotine-Marussi Symposium on Mathematical Geodesy, International Association of Geodesy Symposia, 142:213–219. Springer, Cham.

Gatti A., Reguzzoni M., Venuti G. (2013). *The height datum problem and the role of satellite gravity models*. Journal of Geodesy, 87(1):15–22.

Gerlach C., Fecher T. (2012). *Approximations of the GOCE error variance-covariance matrix for least-squares estimation of height datum offsets*. Journal of Geodetic Science, 2(4):247–256.

Gilardoni M., Reguzzoni M., Sampietro D., Sansò F. (2013). *Combining EGM2008 with GOCE gravity models*. Bollettino di Geofisica Teorica ed Applicata, 54(4):285–302.

Gilardoni M., Reguzzoni M., Sampietro D. (2016). *GECO: a global gravity model by locally combining GOCE data and EGM2008*. Studia Geophysica et Geodaetica, 60(2):228–247.

Hirt C., Kuhn M. (2012). *Evaluation of high-degree series expansions of the topographic potential to higher-order powers*. Journal of Geophysical Research: Solid Earth, 117, B12407.

Mayer-Gürr, T., Rieser D., Höck E., Brockmann J.M., Schuh W.-D., Krasbutter I., Kusche J., Maier S., Krauss S., Hausleitner W., Baur O., Jäggi A., Meyer U., Prange L., Pail R., Fechner T., Gruber, T. (2012). *The new combined satellite only model GOCO03s*. Presentation at International Symposium on Gravity, Geoid and Height Systems GGHS 2012, Venice, Italy.

Mayer-Gürr T., Pail R., Gruber T., Fecher T., Rexer M., Schuh W.-D., Kusche J., Brockmann J.-M., Rieser D., Zehentner N., Kvas A., Klinger B., Baur O., Höck E., Krauss S., Jäggi A. (2015). *The combined satellite gravity field model GOCO05s*. Presentation at EGU 2015, Vienna, Austria.

Pail R., Goiginger H., Schuh W.-D., Höck E., Brockmann J.M., Fecher T., Gruber T., Mayer-Gürr T., Kusche J., Jäggi A., Rieser D. (2010). *Combined satellite gravity field model GOCO01S derived from GOCE and GRACE*. Geophysical Research Letters, 37(20):L20314.

Pail R., Bruinsma S., Migliaccio F., Förste C., Goiginger H., Schuh W.-D., Höck E., Reguzzoni M., Brockmann J.M., Abrikosov O., Veicherts M., Fecher T., Mayrhofer R., Krasbutter I., Sansò F., Tscherning C.C. (2011). *First GOCE gravity field models derived by three different approaches*. Journal of Geodesy, 85(11):819–843.

Pavlis N.K., Holmes S.A., Kenyon S.C., Factor J.K. (2012). *The development and evaluation of the Earth Gravitational Model 2008 (EGM2008)*. Journal of Geophysical Research: Solid Earth, 117(B4):B04406.

Pavlis N.K., Holmes S.A., Kenyon S.C., Factor J.K. (2013). *Correction to "The development and valuation of the Earth Gravitational Model 2008 (EGM2008)"*. Journal of Geophysical Research: Solid Earth, 118(5):2633.

Rapp, R.H. (1989). *Combination of Satellite, Altimetric and Terrestrial Gravity Data*. In: Sansò F., Rummel R. (Eds.), Theory of Satellite Geodesy and Gravity Field Determination, Lecture Notes in Earth Sciences, 25. Springer-Verlag, Berlin, Heidelberg, pp. 261–284.

Reguzzoni M., Sansò F. (2012). *On the combination of high-resolution and satellite-only global gravity models*. Journal of Geodesy, 86(6):393–408.
Reguzzoni M., Venuti G. (2018). *Local versus global height datum adjustment*. Presentation at the IX Hotine-Marussi Symposium, Rome, June 2018.
Reguzzoni M., Venuti G., de Lacy M.C., Carrion D., Barzaghi R., Borque M.J., Gil A.J., Vaquero P.A. (2018). *The use of GNSS/levelling and gravity data for the Spanish height system unification*. In: Vergos G.S., Pail R., Barzaghi R. (eds), *International Symposium on Gravity, Geoid and Height Systems 2016*, International Association of Geodesy Symposia, 148, Springer, Cham.
Reigber C., Jochmann H., Wünsch J., Petrovic S., Schwintzer P., Barthelmes F., Neumayer K.-H., König R., Förste C., Balmino G., Biancale R., Lemoine J.-M., Loyer S., Perosanz F. (2004). *Earth Gravity Field and Seasonal Variability from CHAMP*. In: Reigber C., Lühr H., Schwintzer P., Wickert J. (Eds.), Earth Observation with CHAMP-Results from Three Years in Orbit. Springer, Berlin, Heidelberg, pp. 25–30.
Sansò F., Sideris M.G. (2013). *Geoid determination: Theory and methods*. Lecture Notes in Earth System Sciences, Vol. 110. Springer-Verlag, Berlin, Heidelberg.
Tapley B.D., Bettadpur S., Watkins M., Reigber C. (2004). *The gravity recovery and climate experiment: Mission overview and early results*. Geophysical Research Letters, 31(9):L09607.

Index

F. Sansò et al., *Geodetic Heights*, Springer Geophysics,
https://doi.org/10.1007/978-3-030-10454-2

MIX
Papier aus verantwortungsvollen Quellen
Paper from responsible sources
FSC® C105338

If you have any concerns about our products,
you can contact us on
ProductSafety@springernature.com

In case Publisher is established outside the EU,
the EU authorized representative is:
Springer Nature Customer Service Center GmbH
Europaplatz 3, 69115 Heidelberg, Germany

Printed by Libri Plureos GmbH
in Hamburg, Germany